SMART METERING TECHNOLOGY AND SERVICES – INSPIRATIONS FOR ENERGY UTILITIES

Edited by **Moustafa Eissa**

Smart Metering Technology and Services – Inspirations for Energy Utilities
http://dx.doi.org/10.5772/61356
Edited by Moustafa Eissa

Contributors

Yue Yang, Mahmoud Amin, Carlos Alberto Fróes Lima, Vasiliki Klonari, Jean-François Toubeau, Jacques Lobry, François Vallée, Trong Nghia Le, Wen-Long Chin, Moustafa Eissa, Martin Anda

Published by InTech

Janeza Trdine 9, 51000 Rijeka, Croatia

Publishing Process Manager Edi Lipovic

Technical Editor InTech DTP team

Cover InTech Design team

First published June, 2016

Printed in Croatia

Additional hard copies can be obtained from orders@intechopen.com

Smart Metering Technology and Services – Inspirations for Energy Utilities, Edited by Moustafa Eissa

p. cm.

Print ISBN 978-953-51-2451-1
Online ISBN 978-953-51-2452-8

For EU product safety concerns:
IN TECH d.o.o., Prolaz Marije Krucifikse Kozulić 3, 51000 Rijeka, Croatia, info@intechopen.com or visit our website at intechopen.com.

Meet the editor

Moustafa M. Eissa is a Professor in the Department of Electrical Engineering at Helwan University, Helwan, Egypt. He is a senior member of the IEEE-PES and a recipient of the Helwan University Research Award for 2005 and 2011. He also owns the ETRERA 2020 Prize in the category of Smart Grids (2014) from Europe. He is a recipient of Research Excellence Prize from Ministry of Higher Education and State Encouragement prize in the advanced technology science from the Academy of Scientific Research and Technology for 2002 and 2012 (Egypt). He has researched, consulted and authored many articles in IEEE Transaction on power delivery, IEEE Transaction on Smart Grid, IET, Elsevier and ETEP journals. He has many large-scale grants in the smart meters-based GIS, microgrid and hybrid RES, hybrid energy efficiency with optimization, WAMs-based HVDC control/protection, energy policy makers, DSM- and EMS-based smart grid.

Contents

Preface IX

Section 1 Demand Response and Market Regulation 1

Chapter 1 **Introductory Chapter: Demand Response Incentive Program (DRIP) with Advanced Metering and ECHONET 3**
Moustafa M. Eissa

Chapter 2 **Smart Grid Implementation in Brazil Must Focus on Consumer Behavior and Markets, Regulation, and Energetic Mix Availability 15**
Carlos Alberto Fróes Lima

Section 2 Smart Metering Architecture, Implementation and Application 35

Chapter 3 **Advanced Metering Infrastructure Based on Smart Meters in Smart Grid 37**
Trong Nghia Le, Wen-Long Chin, Dang Khoa Truong and Tran Hiep Nguyen

Chapter 4 **The Post Carbon City and Smart Metering 63**

Chapter 5 **MAC Protocol Design for Smart Meter Network 85**
Yue Yang

Section 3 Platforms for Renewable Resources and Electronics in Smart Systems 105

Chapter 6 **Power Electronics Platforms for Grid-Tied Smart Buildings 107**
Mahmoud Amin

Chapter 7 **Estimating the Photovoltaic Hosting Capacity of a Low Voltage Feeder Using Smart Meters' Measurements 131**
Vasiliki Klonari, Jean-François Toubeau, Jacques Lobry and François Vallée

Preface

Smart meters are advanced energy meters that support two-way communications compared to a conventional energy meters. Hence, they can measure the energy consumption data of consumers and then transmit added information to the utility companies to support decentralized generation sources and energy storage devices, and bill the customer. Besides, smart meters can receive information about electricity price and commands from utility companies and then deliver them to consumers. Integration of smart meters into power grid involves implementation of a variety of techniques and software, depending on the features that the situation demands. Design of a smart metering system is realized on the requirements of utilities and customers. The smart meters offer businesses and households the roadmap to educate and reduce their energy usage in much greater detail than previously possible, when the meter readings are taken in a short time. They promise to help utilities, to improve the accuracy of billing and cut visits to properties to read meters. They can also collect diagnostic information about the distribution grid, home appliances, and can communicate with other meters in their reach. Moreover, smart meters can be programmed such that, only power consumed from the utility grid is billed while the power consumed from the distributed generation sources or storage devices owned by the customers are not billed. The ultimate goal of smart metering is to allow utility firms to forecast energy usage, to improve their performance on the settlement markets and to match supply and demand more closely. There are many types of devices, systems, and programs that can be used to achieve the demand side management. These managements require participation by customers and ultimately rely on the effectiveness of customer-based programs. There are many information technologies, communication media and protocols that are needed for smart metering infrastructure.

The book contains three sections: Section 1: Demand Response and Market Regulation, Section 2: Smart Metering Architecture, Implementation and Application, and Section 3: Platforms for Renewable Resources and Electronics in Smart Systems. Many authors with academic and industrial expertise in the field of the smart metering technology and services contributed to this book which will increase the knowledge and information in this topic for engineering, academics, and students.

Section 1: Demand Response and Market Regulation

This section describes demand response programs based on real-time signal that have proven to be effective methods for utilities to manage system peaks by controlling customer loads. For commercial consumers, these programs conventionally entailed the direct load control of large appliances at the company such as HVDC systems, large motors, and pumps.

The communication systems applied in transmission or telemetry of data and control will send signals between the meter interface units and the central office. Such communications can take the shape of telephone, power line carrier, radio frequency, or cable television. These system components involved in the communication systems depend on the communication media used. The Demand Response Incentive Program (DRIP) submitted to the Electricity Company (EC) is a response to motivate participants to reduce the load at peak periods between 1.00 and 5.00 pm during working days (Saturday to Wednesday), from the first of June to the end of September, the peak in the Electricity Company. The program is an incentive rate for the customers, which they get as return if they reduce the power demand during the system peak period. Commercial Energy Management System (CEMS) is a system for cleverly managing the various types of energy used in the industry. Installing CEMS makes data on electric power generation and utility usage visible on monitors and other screens, facilitating "control" of CEMS compatible machines and loads. By making the electricity used by load more "visible", each member of the company will become more aware of saving energy, wasteful use of electricity will be eliminated, and energy costs will thus be reduced. Shared communication protocols are necessary to achieve two-way communication between CEMS controllers and various home appliances, household devices, etc. This function is provided by the ECHONET Lite Specification. With CEMS controllers adopting ECHONET Lite and devices compatible with CEMS, it should become possible for different manufacturers' products to be connected together for use. In Brazil, although some critical components are already monitored, such as interface audit meters as well as large customers meters, the data analyses are neither systematic nor in real time for all dealers. In the best of cases, when some kind of accounting science is applied to this information, this knowledge becomes sectored and used to support a business segment. Therefore, sensing, metering, data presentation, their systematic usage, simulation environment, tests, business intelligence reports, as well as electricity quality, need a business (re)organization to go through this focus. An evaluation of the Brazilian market was done, considering the up-to-date international experiences and running an application, specially built to demonstrate the domestic consumption as a case study and the required evolution of systems and strategies to move on to this historical moment of development and reorganization of the energy market as well as the legislation/regulation.

Section 2: Smart Metering Architecture, Implementation and Application

This section deals with the lack of situational awareness, automated analysis, poor visibility, and mechanical switches, today's electric power grid, ageing and ill-suited to the demand for electricity, which has gradually increased in the twenty-first century. Besides, the global climate change and the greenhouse gas emissions on Earth caused by the electricity industries, the growing population, one-way communication, equipment failures, energy storage problems, the capacity limitations of electricity generation, decrease in fossil fuels, and resilience problems put more stress on the existing power grid. Consequently, the smart grid (SG) has emerged to address these challenges. To realize the SG, an advanced metering infrastructure (AMI) based on smart meters is the most important key.

Also, the buildings and districts are an appropriate focus for smart metering infrastructure in the urban environment. While properties and buildings have traditionally been metered for revenue recovery purposes, energy management of these buildings has not been available. In the way the section explains the smart meters as a vital component to the making and management of post-carbon cities and can be used to monitor not only electricity use but also water and gas consumption. Energy management systems combined with structured metering also enable consumers with renewable energy generation such as photovoltaic (PV) panels to monitor their own generation, consumption, import, and export. As battery storage becomes integrated with renewable energy generation, consumers will have the ability to consume cheaper renewable energy than can be bought from the grid and sell energy back to the grid at the most economically viable times. While uncertainty surrounds the grid and its impact on rising electricity prices, smart metering, intelligent control systems and utilities, offering consumers more amenities and the ability for consumers to participate in the wholesale market will ensure that the smart grid can contribute to future carbon neutral urban environments.

The new generation of power metering system - i.e., advanced metering infrastructure (AMI) - is expected to enable remote reading, control, demand response, and other advanced functions, based on the integration of a new two-way communication network, which will be referred to as smart meter network (SMN). In this section, authors focus on the design principles of multiple access control (MAC) protocols for SMN. First, they list several features of SMN relevant to the design choice of the MAC protocols. Next, they introduce some performance evaluation metrics and give a survey of the associated research issues for the SMN-MAC protocols' design. In addition, they also note progress within the new IEEE standardization task group (IEEE 802.11ah TG) currently working to create SMN standards. After that, in order to emphasize the importance of the performance metrics mentioned before, they give several MAC protocol design examples that could solve the associated research issues and challenges for the SMN.

Section 3: Platforms for Renewable Resources and Electronics in Smart Systems

Renewable energy sources (such as sun, wind, water, or fuel cells) are finding great interest for either grid-tied or off-grid arrangements in smart green buildings. It must be either used when generated, stored for future use on site, delivered to the power grid, or shared among a combination of these. Grid-tied buildings are connected to the utility grid service lines. Off-grid buildings have no connection to utility service lines. Both types employ inverters to convert power from direct current (DC) to alternating current (AC), and most off-grid systems have batteries to store energy for use when needed. Accordingly, power electronics systems are playing an important role as the enabling technology for smart grid. In addition, smart meter represents the interface part between the green building and the utility grid. In order to realize the interaction among both systems, a bidirectional power conditioning module is needed. This section introduces the different power electronics platforms suitable for grid-tied smart green buildings (such as residential homes, commercial, and industrial) as well as its integrative functionality with advanced metering Infrastructure (AMI). In order to show the superiority of these platforms in conjunction with smart meters, a hardware case study with one of the most popular power electronics topologies is presented in this section.

Maximizing the share of renewable resources in the electric energy supply is a major challenge in the design of the future energy system. Regarding the low voltage level, the main focus is on the integration of distributed photovoltaic (PV) generation. Nowadays, the lack of monitoring and visibility, combined with the uncoordinated integration of distributed generation, often leads system operators to an impasse. As a matter of fact, the numerous dispersed PV units cause distinct power quality and cost-efficiency problems that restrain the further integration of PV units in LV feeders. The PV hosting capacity is a tool for addressing such power system performance and profitability issues so that the different stakeholders can discuss on a common ground. Photovoltaic hosting capacity of a feeder is the maximum amount of PV generation that can be connected to it without resulting in unacceptable power quality. This section demonstrates the usefulness of smart metering (SM) data in determining the maximum PV hosting capacity of an LV distribution feeder. Basically, the section introduces a probabilistic tool that estimates PV hosting capacity by using customer-specific energy flow data, recorded by SM devices. The probabilistic evaluation and the use of historical SM data yield a reliable estimation that considers the volatile character of distributed generation and loads as well as technical constraints of the network (voltage magnitude, phase unbalance, congestion risk). As a case study, an existing LV feeder in Belgium is analyzed. The feeder is located in an area with high PV penetration and large deployment of SM devices. When the probabilistic character of the network states and of the applied voltage limits is considered, the estimated PV hosting capacity is proved to be much higher than the one obtained with a deterministic approach, based on worst case energy flow profiles.

Prof. Moustafa M. Eissa
Faculty of Engineering at Helwan-Helwan University
Egypt

Section 1

Demand Response and Market Regulation

Chapter 1

Introductory Chapter: Demand Response Incentive Program (DRIP) with Advanced Metering and ECHONET

Moustafa M. Eissa

Additional information is available at the end of the chapter

http://dx.doi.org/10.5772/64206

1. Introduction

The rollout of smart meters plays an important role for the world transition to a low-carbon economy and helps in facing some of the long-term challenges for ensuring an affordable, secure, and sustainable energy supply. Smart meters have the potential to transform consumers' relationship with energy bringing considerable benefits and also for energy industry. A smart metering solution generally delivers a range of applications using an infrastructure comprising networked meters, communication networks, and data collection and management systems. Recently, smart metering and infrastructure using advanced metering infrastructure [1] has become one of the hottest topics within electric power utilities. Smart metering and infrastructure is not a tool to capture customer energy consumption every month or two, but an integrated hardware and software architecture is capable of capturing real-time consumption, demand, voltage, current, and other information. In other words, the system functionality of smart metering and infrastructure goes far beyond just obtaining a monthly meter reading. Establishing bidirectional communication and smart metering technologies that can record customers' load profiles, the smart metering and infrastructure provides utilities with system-wide sensing and measuring capability [2–4].

When establishing smart meters, many areas should be considered, the most important are dynamic tariffs, smart grids, and feed-in tariffs. The dynamic tariffs permit the utilities to charge many rates at different day and focus as "spot market" for business customers. The smart grids are precisely balancing the supply and load for in order to earn the optimum benefit of intermittent RES. The feed-in tariffs authorize the utilities to measure and reward electricity generated on-site by microgeneration technologies. This contains photovoltaic panels (PVs), microcombined heat and power, and small-scale wind turbines. Smart meters offer several additional features of interest to the utility market in comparison with traditional mechanical and electromechanical meters, and smart meters include the following:

- High reliability and hardness
- High accuracy
- Anti-tampering function
- Automation in reading
- Support of nonlinear and low-power factor loads
- Self-calibrated
- Security
- Applying with different programs using advanced billing (time-of-use, prepay, etc.)

Whether the measurements can be gas, water, heat, or electricity, some or all of these features apply and are making the smart meters the solution of choice in both new and existing markets.

- The smart meter can measure the amount of electricity at customer sides. The differences between a smart meter and tradition mechanical meter are that the smart meter automatically can remotely transmit total electric usage to company Power using via radio signal for the customers.
- The smart meters are considered as a part of integrated program that will pay for itself through reduced theft of electricity, operational efficiencies, and energy savings.
- Several technologies are currently in use to achieve automated meter reading of electronic meters or to retrofit existing mechanical/electromechanical meters.

Electronic meters can read and communicate automatically through different mechanisms such as:

- Short-range infrared
- Short and long radio frequency range
- Broadband
- Telephone line data modem
- Short to medium power line carrier range
- RS-485 serial port

The automated reading advantages that they can be obtained using communicating with a handheld device (RF, infrared—up to several hundred feet away). This of course cannot eliminate the need for operators who visit locations, and in such case, the readings are accurate and speed up the process. The smart meters data flow can be given in **Figure 1**.

The communications systems applied in transmission, or telemetry, of data, and control will send signals between the meter interface units and the central office. Such communications can take the shape of telephone, power line carrier, radio frequency, or cable television. These system components involved in the communications system depend on the communication media used.

There are two new billing technologies that are made possible by the implementation of electronic meters: time-of-use and prepay.

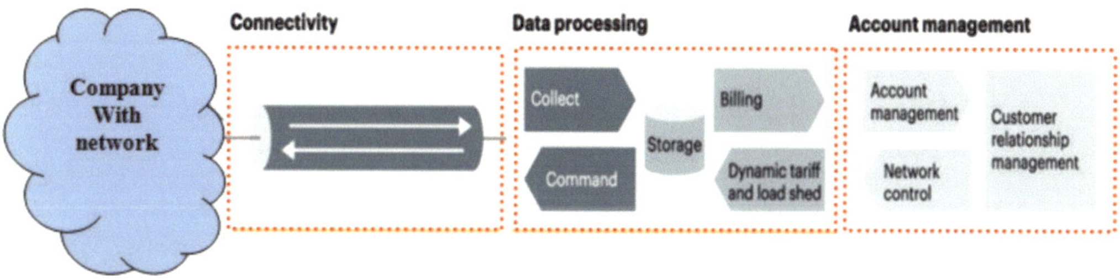

Figure 1. The smart meters data flow.

1.1. Time-of-use

The TOU pricing can also be used as an involuntary way for consumers to adjust the electricity consumption in different time axis in accordance with the cost of electricity. Time-of-use defines as the different tariffs for the use of the utility at different hours of the day or day of the week. This technology helps the utilities to shape the demand in order to optimize the utilization of the available capacity throughout the day.

1.2. Prepay

- The technology of prepay is applied to reduce the financial cost of payment of the utilities. The user is permitted to purchase certain amounts of the service ahead of time and receives credits that are charged on smart or magnetic cards. The meter coupled with a card reader and acts as a gate controlling the delivery of the electricity service.

Automatic meter reading system (AMR) is the remote collection of consumption data from customers' utility such as electric meters using radio frequency, telephony, power line, or satellite communication technologies, and processes the data to generate the bill.

For their advanced metering infrastructure (AMI), utilities require an economical solution to enable fast and secure communications to smart meters, to allow new market-driven billing structures to be implemented.

- Now there are different kinds of communication and protocol types that used smart metering. A combination of GPRS and power line carrier communication is used in many countries such as Italy, Denmark, and Finland. These European examples of the PLC technology are used between the smart meters and data concentrators, and GPRS is used between the concentrator and gateway to the data management system.

2. Demand response

Demand response is the process of scheduling the loads to reduce the electric energy consumption and or the maximum demand. It is basically optimizing the processes and loads to improve the system load factor [5].

The concept of the demand response is traditionally used as load management methods for changes in the equipment and consumption patterns on the customer side.

The load management methods can be applied by an industry or a utility. Typical actions can be given such as load shedding and restoring, load shifting, achieving energy efficient programs and equipment, energy storage, cogeneration and nonconventional energy sources, and reactive power control.

The load management methods provide customers with the option to avoid or curtail central electricity during the peak hours. Some mechanisms for load management programs are given in [6, 7].

The major benefits of demand response is the reduction in maximum demand, reduction in power loss, better equipment utilization, and saving through reduced maximum demand charges. Such a program of load shifting to the off-period is to reduce the demand in the peak period by shifting some appliances and equipment to the off-peak periods. It is important to focus about the costs arise that showed it can be possible to take features of incentives and favorable pricing gained by utilities in order to encourage consumers to use energy during off-peak period. In this case, it enables the utility to change load patterns.

Demand response is considered as an essential tool for the electric utility in facing the growing demand for electricity. Demand response is a subset of wider category of consumers for energy solutions known as demand-side management (DSM). Beside the demand response, DSM includes energy efficiency programs and conservation. Demand response defines as changes in the electric usage by end uses from their regular consumption patterns in response to changes in the price of electricity over time, or it can be incentive payments designed to produce lower electricity use at times of high wholesale market prices [8].

Classification of the demand response programs as follows:

- Demand response based on real-time pricing (RTP), critical-peak pricing (CPP), and time-of-use (TOU) tariffs that facilitate customers time-varying rates that reflect the value and cost of electricity in different time periods.

- Demand response based on incentive programs can help in participating customers to reduce their loads at times requested by the program sponsor, issued either by a grid reliability problem or by high electricity prices.

To encourage reduction in peak demand, many utilities have already implemented time-of-use rates (TOU) or have plans for introducing such rates [9, 10].

The purpose of utilizing these programs is to reduce the load curve during the peak periods. However, some utilities applied these programs mandatory during the peak periods. During the day, the programs can be applied successful in case of having the customers peak demands.

2.1. Time-of-use-tariff

- Time-of-use defines as various tariffs for the use of the same utility at different intervals of hours per day or day of the week. This technology helps the utility companies to shape the demand in order to optimize the utilization of the available capacity through the day.
- With high rates, it can charge during the peak hours and incentive the user to make a more rational efficient use of the resources.
- Smart meters can incorporate inexpensive real-time clock (RTC) and calendar (RTCC) circuitry to keep track of utility usage in real time.

2.2. Interruptible load tariffs

These incentive rates for the customers, which they get if they interrupt or reduce the power demand during the system peak period or emergency condition. Large industrial consumers can use interruptible tariffs. The implementation of interruptible tariff structure involves unbundling electric service and offers the customers a range of rate reliability choices. Load contract with utility is signed for an interruptible to reduce their demand as and when requested by the utility.

2.3. Dynamic pricing (dispatchable rates)

Different prices during different time periods so-called retail prices for energy consumed offer and reflect the fact that power generation costs and wholesale power purchase costs vary during different time periods. There are different types such as dynamic versions of time-of-use pricing, critical peak pricing, and real-time pricing.

3. Proposed program

Different terms in use in the demand side related to each other but with slightly different focuses. These definitions are as follows:

- Demand-side management (DSM): The activities of the utility influence use of the customer for electricity. This includes the planning, implementation, and monitoring of many activities. Such activities are designed to incentive consumers to change the consumer patterns of usage.

- Demand response (DR): This type of program used as mechanisms to manage the demand response to supply conditions in signal communication.
- Demand-side participation: In the competitive electricity market, a set of strategies can be applied via customers to contribute to economic, system security, and environmental benefits.

The demand response program is a potential resource for satisfying the nation's energy needs. If the peak demand for energy is lowered, the demand response programs will reduce constructing new plants and avoiding the expensive generation units. The Federal Energy Regulatory Commission (FERC or Commission) staff reports—*A National Assessment of Demand Response Potential* (National Assessment), submitted to Congress in June 2009—current demand response programs tap less than a quarter of the total market potential for demand response [11]. The running programs have missed a potential portion of the cost-effective demand response significant, and for this reason, the action needs to be taken to either create new programs or apply existing ones with cost-effective.

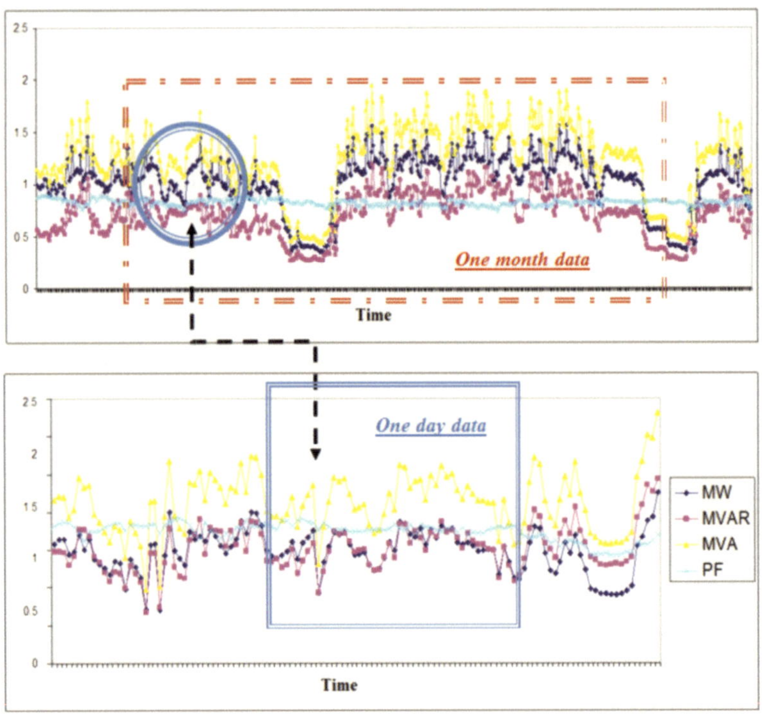

Figure 2. The load curve used for calculating the base power in case of nonavailability of the recorded data.

The program of **DRIP**: Demand response incentive program submitted to the electricity company (EC) is a response to a voluntary to motivate participants to reduce the load at peak periods between one o'clock noon until five o'clock pm during working days (Saturday to Wednesday), from the first of June until the end of September, the peak in the electricity

http://dx.doi.org/10.5772/64206

company. The program is an incentive rates for the customers, which they get a return if they reduce the power demand during the system peak period. The electricity company (EC) will pay for the customers an incentive which depends on the level of reduction in demand power each month during the peak periods. One of the main objectives of the program (**DRIP**) is to raise the efficiency of energy demand and further secure the electrical grid. The participants can be in cooperation with the electricity company to achieve the program of **DRIP** to reduce loads through redistribution of electrical loads, the use of reserve generation, the use of storage cooling, shifting electrical loads, etc. **Figure 2** shows the load curve used for calculating the base power in case of nonavailability of the recorded data.

The procedures of the **DRIP** program involve changes to load consumption pattern on the customers' side. To achieve the program, the following three methods should be satisfied:

3.1. Remote load control method

The program can be achieved by using the control of electric load (load redistribution or stand-by electricity generation). This can be done through switching load off at this time or through the use of on-site generation for part of the load.

3.2. Obligated reduced power

The customers should introduce application form to indicate the level of electricity demand they will consume (in MW) during the delivery period. This is the obligated level

3.3. Payment procedure

Payment will be made after the meter has been read and some calculations. The payment calculation can be classified as:

a. ***A—Power reduced payment (PRP)***

b. ***B—Demand incentive payment (DIP)***

c. ***C—Violation payment (VP)***

Figure 1 shows the load curve used for calculating the base power in case of nonavailability of the recorded data.

4. Outage detection

In the smart meters scenario, the utilities are familiar by the power outages at customer location (obligated level). Using the smart meters data and obligated level of contracts, you can easily identify the level of reduction and amount of outage as being less or large. So, with smart grid, the smart meter can detect power outage and inform back in real time, providing the outage location and lot more technical data than what the service personnel typically had.

5. Service connect/disconnect

The smart meters have the capability to connect a customer service or disconnect avoiding many of problems for the utilities and customers. The customers should precisely identify the controlled load under switching with the utility during the obligated contracted period. This facility greatly improves customer satisfaction, by being able to respond to a request 'online', without having to set up appointment for field visit.

6. Peak demand reduction

The utilities bear maximum costs in estimating and meeting the peak demand of consumers, using peaking plants and providing very higher amount than estimated peak. Smart grid using smart meters can offer many methods to flatten the peak demand, such as:

a. Issue an awareness of the increasing demand

b. Implement incentive rate for reducing the peak usage compared to off-peak

c. Implement active ***DRIP*** to tune specific energy devices (reducing load settings) into 'load shedding'

d. Increase insight into consumer use patterns to the Utility

6.1. Support for *DRIP* rates

a. AMI enables two-way flow of information between the meters and the utility, enabling active monitoring of energy usage, and convey and help implement ***DRIP*** rates. Smart grid strives to flatten the peak demand, and implementing ***DRIP*** rates is a good way to reduce peak demand. In some utilities, critical peak rates can be up to 100 times the off-peak rates, leading the consumer to be 'more aware' of the difference in 'energy supply' conditions.

7. Commercial energy management system (CEMS)

Commercial energy management system (CEMS) is a system for cleverly managing the various types of energy used in the industry. Installing CEMS makes data on electric power generation and utility usage visible on monitors and other screens, facilitating "control" of CEMS compatible machines and loads. By making the electricity used by load more "visible," each member of company will become more aware of saving energy, and wasteful use of electricity will be eliminated and energy costs will thus be reduced. And when you would reduce reduction, you will be able to "control" energy use by switching off or some of CEMS compatible loads with a single action, thus avoiding energy wastage. The management process can be done by actually study well the loads and period of peaks. **Figure 3** shows the ECHONET is a communication protocol design.

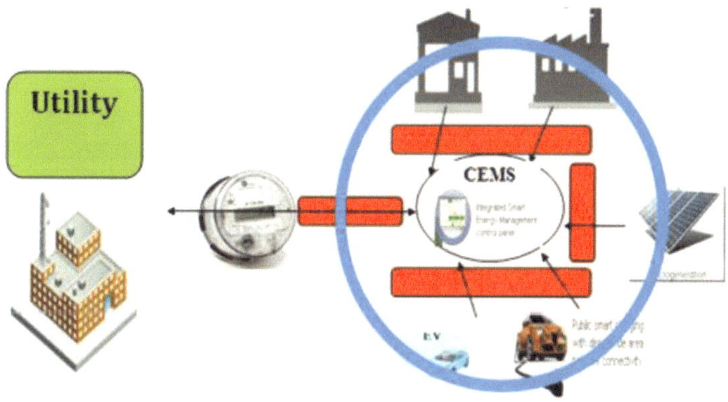

Figure 3. ECHONET is a communication protocol design.

8. Communication protocol

Smart electric energy meters digitally measure electricity usage and automatically send measured data to power companies. They can also communicate with CEMS controllers via ECHONET Lite and express electricity usage status in visible form, based on the meter readings used for billing [12].

9. ECHONET Lite: a communication protocol supporting CEMS

Shared communication protocols are necessary to achieve two-way communication between CEMS controllers and various home appliances, household devices, etc. This function is provided by the ECHONET Lite Specification. With CEMS controllers adopting ECHONET Lite and devices compatible with CEMS, it should become possible for different manufacturers' products to be connected together for use [12].

ECHONET is a communication protocol designed to create the "smart commercial" of the future. Today, with Wi-Fi and other wireless networks readily available in ordinary industry, there is a growing demand for load inside the factory to be controlled using smart phones or controllers, or for electricity usage to be monitored in order to avoid wasting energy [12].

To achieve this kind of low-energy, comfortable, safe and reassuring lifestyle, we first need to create a system of rules or a "communication protocol" that can be read by any manufacturer's equipment. This is where ECHONET comes in.

The ECHONET Lite specification, in particular, is a communication protocol compatible with the now ubiquitous Internet. It is designed for ease of use and is simpler than the ECHONET specification. The ECHONET Lite specification is already compatible with more than 100 types of device and is also being adopted by the smart electric energy meters that will be installed in all households in future. **Figure 4** shows the overall structure of the proposed idea [12-14].

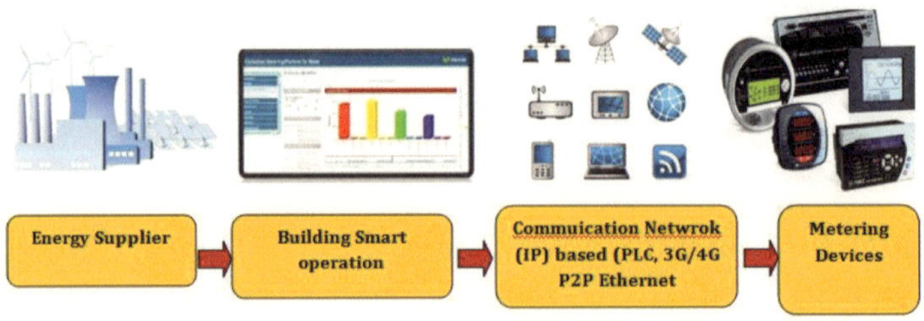

Figure 4. Flow chart for smart metering operations.

10. ECHONET lite system architecture

ECHONET Lite System Architecture This section specifies the ECHONET Lite system configuration and system architecture. **Figure 5** shows the system architecture. An ECHONET Lite system incorporates many devices with the same properties, security, management, etc. So, the major part that ECHONET Lite can manage is referred to as a domain. A domain will be specified as the range of controlled resources (company load, appliances and motors,

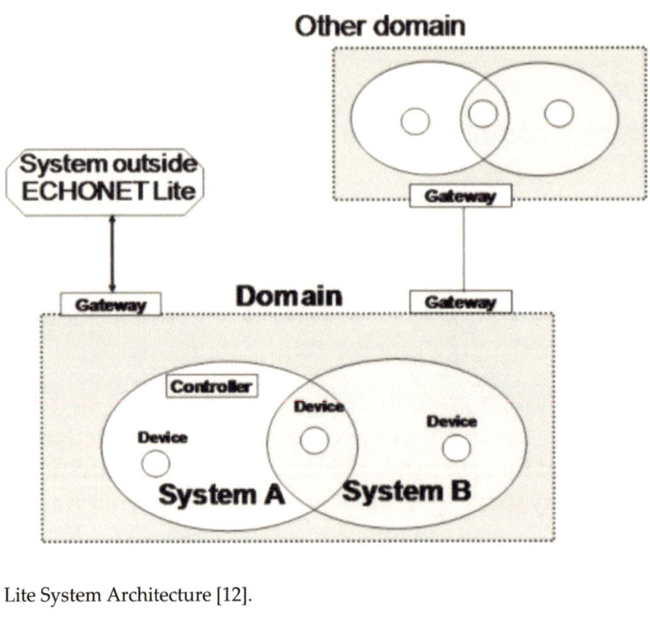

Figure 5. ECHONET Lite System Architecture [12].

sensors, controllers, control, etc.) present within the network range determined by ECHONET Lite. A system is defined as which performs communication and linked operations between devices and the controllers that control, monitor, and operate them and among devices themselves [12].

Author details

Moustafa M. Eissa

Address all correspondence to: mmmeissa@yahoo.com

Faculty of Engineering, Helwan University at Helwan, Cairo, Egypt

References

[1] David Bakken, ed. 2014. Smart grids clouds, communications, open source, and automation, Washington State University, Taylor & Francis group, LLC, CRC press, Informa business.

[2] Advanced Metering Infrastructure, National Energy Technology Laboratory. 2008. https://www.smartgrid.gov/files/advanced_metering_infrastructure_02-2008.pdf

[3] Smart Meter 2012, Chartwell Inc. 2012. https://www.chartwellinc.com/smart-meter-opt out programs-spread-across-north-america-new-chartwell-report-details-various-efforts/

[4] AMI Technology Trials Report, Department of Primary Industries. 2007. Australia.

[5] Isaksen, L., Ma, F.S., and Simons, N.W. 1981. Bibliography and load management. IEEE Transactions on Power Apparatus and Systems. PAS-100(5), 2597–2599.

[6] Ashok, S., and Banerjee, R. 2000. Load-management applications for the industrial sector. Applied Energy 66, 105–111.

[7] Effler, L., Schellsfede, G., and Wagncr, H. 1992. Optimization of energy requirement and load management. IEEE Transactions on Power Systems 17(1), 327–333.

[8] Jaffe, A.B., Stavins, R.N. 1994. The energy-efficiency gap. Energy Policy 22(10), 804–810.

[9] Apolinario, I., Felizardo, N., Leite Garcia, A., Oliveira, P., Trindade, A., Vasconcelos J., and Verdelho, P. 2004. Application of additive tariffs in the electricity sector. WEC Regional Energy Forum-FOREN 2004.

[10] Sheen, J.N. 1994. TOU pricing of electricity for load management in Taiwan power company. IEEE Transactions on Power Systems (9), 388–396.

[11] A National Assessment of Demand Response Potential. (June 2009). Available at: http://www.ferc.gov/legal/staff-reports/06-09-demand-response.pdf (National Assessment).

[12] APAN 39th Conference IPv6 Workgroup Session, The open standard technology deployment on a smart meter and a mobile network in Japan. 2015. http://archive.apan.net/meetings/Fukuoka2015/Sessions/23/Excerpt_smart-house_150305_1@APAN.pdf

[13] A report to the United States congress. Benefits of demand response in electricity markets and recommendations for achieving. http://www.electricity.doe.gov/documents/congress_1252d.pdf

[14] Nelson, S.K., and Hobbs, B.F. 1992. Screening DSM programs with a value based test. IEEE Transactions on Power Systems 7(3), 1031–1043.

[15] U.S. Department of Energy. Benefits of demand response in electricity markets and recommendations for achieving them. A Report to the United States Congress Pursuant to Section 1252 of the Energy Policy Act of 2005. February 2006 (February DOE EPAct Report). http://www.oe.energy.gov.

Chapter 2

Smart Grid Implementation in Brazil Must Focus on Consumer Behavior and Markets, Regulation, and Energetic Mix Availability

Carlos Alberto Fróes Lima

Additional information is available at the end of the chapter

http://dx.doi.org/10.5772/62582

Abstract

In Brazil, although some critical components are already monitored, such as interface audit meters as well as large customers meters, the data analyses are not systematic nor in real time for all dealers. In the best of cases, when some kind of accounting science is applied to this information, this knowledge becomes sectored and used to support a business segment.

Therefore, sensing, metering, data presentation, and their systematic usage, including simulation environment, tests, business intelligence reports, and electricity quality, need a business (re)organization to go through this focus.

An evaluation of the Brazilian market was done, considering the up-to-date international experiences and running local pilots, specially built to demonstrate the domestic consumption, as a case studies, and the required evolution of systems and strategies to move on to this historical moment of development and reorganization of the energy market as well as the legislation/regulation. Metering possibilities are presented to improve smart grid implementation goals, recognizing customers' profiles, promoting innovative products and business.

Keywords: consumer, energy efficiency, energy planning, metering, smart grid

1. Introduction

The smart grid technologies present themselves as opportunities to create new energy business. All stakeholders must be involved with organizing, building, and upgrading the power grid in its aspects of quality, availability, infrastructure, standards, reliability, interconnectivity, and sustainability. It is necessary the awareness of supply-demand target, incomes, and a strongly understanding of clients. The governmental strategies must be clear, with regulatory and legislative initiatives to foster new business and protect the public interest. Consumers (as clients) must be heard, as they become active players in the energy market. As far as they develop a dynamic relationship with the operating power industry, new conditions and requirements need to be created in order to lead the strategic transformation inside and outside regional businesses.

Deals on new energy sources, new technologies, new possibilities of differentials service and prices have been studied. Around the world, incentives as well as the evolution of regulation rules have been seen as fundamental roles to maintain and to expand the power supply and demand-side management, with implications for a better relationship between client-consumers, dealers, and incomes/revenues on invested capital. It also has been important to relate and to rethink the affordability of the tariffs and energy delivery costs to clients.

The way worldwide smart grid solutions are interrelating with the energy sector, resulting in the enterprise transformation, is not working in the same way in Brazil energy sector organization. Regional and social improvement or adaptation are expected but current monopoly way to offer energy must be transformed into a client provider relationship, aggregating new products/services to the sector portfolio to start the (re)evolution. A national energy modernization planning must be established in the near future, pointing out renewable and distributed energy generation as services possibilities, and large-scale smart metering and home metering deployment as efficient tools to achieve information and energy consumption habits modification. Standardization and equipment appropriation (costly and technically) appropriated are expected soon.

Evolving energy business in Brazil does not appear in this analysis merely as a possibility but as fact to be accomplished and a business to be investigated. The governmental strategies must be clear, with regulatory and legislative initiatives to foster new business and protect the public interest, yet considering the climate and the unique energetic mix of the country. Questions related to investments and their profits should be answered according to the regional energy business, as well as the consumers' participation and obviously new legislation and market regulation. Services and products to be offered by the Brazilian energy dealers should evolve correspondingly in order to improve business and recognition as energy solution providers.

Consumers (as clients) must be heard, as they become active players in the energy market. As they develop a dynamic relationship with the operating power industry, new conditions and requirements need to be created. A level of commitment to an energy efficient and sustainable model will depend on a number of rearrangements at the Brazilian energy industry and country governmental guidelines, such as the adequacy structure of energy generation and delivery, regulation, and standardization reinforcement, and also the understanding of customers' needs and socio-cultural efforts to motivate the conscious use of energy.

In the following, some scenarios will be discussed, based on the up-to-date international experiences and running local pilots, especially built to demonstrate the domestic consump-

http://dx.doi.org/10.5772/62582

tion, as a case studies. These implemented cases are presented in order to demonstrate the domestic consumer impact to the energy use as well as the required evolution of systems and strategies to move on to this historical moment of development and reorganization of the energy market as well as the legislation/regulation. The local energy industry should forecast strategies to evolve to this new market while been considered as an energy and services provider. Metering possibilities are presented herein to improve smart grid implementation goals. The Brazilian Regulator with the Governmental support must be in charge and promote the essential of smart grid directions to obtain the economic and social advantages as the rest of the world have been exercising.

2. Worldwide smart grid context as reference

Smart grid has a role on developing new business opportunities, as technically and strategically presents itself as the renewal of the world power industry. Its reflexes, mainly in emerging countries, where it should and can guarantee the presence of these countries as economic powers in the near future, or influence their tendencies toward development. Following this path of changes, one can find the structural and operating conditions for the evolution of the Brazilian power grids, forecasting possibilities and issues which reflect a different reality from those in European, American, and Japanese markets.

In this way, the knowledge of the structures, which guide the business, must be investigated and organized from the generation until the effective delivery of electricity to the customer/end user. This paper also sought to analyze a more effective participation of this new agent, the customer, who can also be a generator and active participant in the electric grid.

A wider offer, with the availability of associated resources/services, must go through a cultural change in relationships and new commitments, as well as being regulated and motivated by present public benefits associated to this new setting. A new perspective of regulating entities must be established in order to start the transformation process proposed by smart grid and to validate the transition of a network in obsolescence into a controlled, controllable, and guaranteed business space for all stakeholders.

The need for electricity, fuels, telecommunications, and/or water utilities to take part in this process reflects the concern with the possibilities and demands of this change, in a broad sense, involving both political and social movements. The necessary effective participation of water supply companies in the Brazilian market reflects the characteristics of the national electricity matrix based on hydroelectric generation. Different from Europe, USA, and Japan, gas companies are only now starting to take over their place in thermo-electrical generation in Brazil, with a differentiated representation in the process and on the influence over operational treaties.

The techno and operational smart grid moment in Brazil is based on proving the concept, adapting models. and testing these models to regional reality. Due to the large social and cultural diversity, the great territorial extent and the differentiated demand for energy by region, this adaptation is of utmost importance. This is necessary as both an opportunity and a challenge in the adhesion to the efficiency and updating of energy business structures. Offerings of new energy sources, new technologies, new service possibilities, and different pricing should be carefully studied. The structuring of incentives and the evolution of regulating devices are fundamental to the maintenance/broadening of the energy offering, and of commitment possibilities between client-customer, dealers,[1] and return on invested capital.

Many opportunities are presented by the characteristics of current electricity grids, whose energy delivery systems [1] are almost entirely mechanical, with a modest usage of sensors, minimal electronic communication and usually with no electronic control. Electricity companies, following the trend of other industries, must update themselves with the use of sensors, communication, and computational skills to expand the overall functionality of supplying electricity, controlling and through feedback, continuously self-adjusting.

This technological gap and apparent simplicity in presenting the evolution as a change to the digital environment can be translated, however, in a multitude of possibilities, broadened by the questionings of energy usage and climate change, poised from COP-16 [2]. COP-21 reinforced all the world efforts to the conscious energy generation and consumption. These possibilities bring along business variables that need to be researched, and mainly, dynamically integrated in the future business moment [3]: new energy sources and electricity generation, storage, transmission, distribution, electric cars, distributed resources, voltage distribution practices, consumption, demand, and end-user commitments, reliability, energy usage optimization, mitigation of environmental impact, and also energy industries assets management, controls and costs (return on investments). Other variables, with more subjective connotations than those presented at this moment, such as welfare user commitment and customer relationship must also be considered and listed to measure the general impact on the planning of changes.

It is expected that with Smart grid as an advanced system, there will be a productivity increased, with consequent repercussion in the electricity usage. At the same time, it is expected that smart grid will organize the backbone for implementing new technologies in the future. In Brazil and other countries under developing economies, this conscious positioning strategically assures the guarantee of necessary energy conditions for future growth that they have been preparing themselves.

Some actions and samples are presented herein on the implementation of smart grids from projects and results of actions to promote grids development and business intelligence. Certain interest points are highlighted and should be used as reference for a Brazilian new electricity business modeling:

[1] In Brazil, the energy business has been regulated ever since 1995 (Concessions Act – Law # 8987 from February 13th, 1995), with the market being formatted through the concession to private consortia for exploiting the market of either in electricity generation, transmission and distribution/trading. Some Brazilian regions are still served by government electricity companies due to regional low performance operation situations, which so far have made it impossible for them to be privatized. In December 1996, Law #9427 created Agência Nacional de Energia Elétrica (ANEEL) as regulatory agency for the sector.

http://dx.doi.org/10.5772/62582

- United Kingdom: the British structure to customer choice, with implementation of intelligence for the electricity offering as a free market (where the customer could elect its energy supplier), places England one step ahead in the restructuring of the energy business [4]. Their energetic mix, the conception of sensing into their grids, their customer care and services associated offerings, yield an apparatus that should allow, without traumas, the (re)evolution of their electricity services and transition to Smart grid. They are still discussing the customer data organization. Consumers will expect to have a choice of tools for viewing and managing home energy use. In the English model, the cost and acquisition of meters are customer responsibility;
- Japan: the country has a well-designed plan, which is being applied, for the efficiency of equipment, appliances, household gadgets, buildings, transport, and industrial production, with established targets and regulations [5, 6];
- United States: there were several test fields for the Smart grid concepts and definitive installation, and these must be evaluated in their achievements and as learning centers. Traditional and emerging forms of electricity on site generation had evaluated in some residences and verified the efficiency of this type of grid improvement [7];
- European Community: several projects have been implemented, as tests and even in national levels [8].

As presented, the consumption and growth environment needs to be understood to improve the grid and the customer relationship. Considering the grids digitization heralded by smart grid, sensing and metering represent a substantial possibility of significant change in the relationship between the distributor and the consumer, capable of assembling demand management mechanisms. However, it means an investment of high figures considering the amount of consumers that shall be assisted. These sensors/meters must be integrated through a close to real-time communication system. Data must be managed through a fast simulation system and computational modeling capacity, being presented in order to empower both operators and managers.

At this point, it is necessary to prove the concept, adapt models, and test these models to the regional reality. The structuring of incentives and the evolution of regulating devices are fundamental to the maintenance/broadening of the energy offering, and of commitment possibilities between client-customer, dealers[2] and return on invested capital.

3. Overview of functional structure of a smart grid operation

Considering the current stage of technology at the electricity companies, mainly in Brazil, from the basic concept of its distribution grids up to the operational organization of its business

[2] In Brazil, the energy business has been regulated ever since 1995 (Concessions Act – Law # 8987 from February 13th, 1995), with the market being formatted through the concession to private consortia for exploiting the market of either in electricity generation, transmission and distribution/trading. Some Brazilian regions are still served by government electricity companies due to regional low performance operation situations, which so far have made it impossible for them to be privatized. In December, 1996, Law #9427 created Agência Nacional de Energia Elétrica (ANEEL) as regulatory agency for the sector.

through billing, many changes, systems, and cultural transformations that smart grid concepts should improve and smart metering will be the vehicle. Smart metering needs a smart telecommunication and data infrastructure to be useful as innovative business solution and could improve products and services offering. This paradigm of consumer recognition, privacy information security, and market (profits) possibilities will evolve. Anyway, functionalities and commitments are highlighted, which can be grouped as:

Visualization of real time energy system: the grid sensing (metering) is a relevant item for the electricity system, broadening the knowledge over the grid and supporting critical components operation. These sensors must be integrated through a real-time communication system. Data must be managed through a fast simulation system and computational modeling capacity, being presented in order to empower both operators and managers and improving grid self-healing.

Sensing, data presentation, their systematic usage, simulation environment, tests, business intelligence reports, as well as electricity quality need a business (re)organization to go through a well-defined focus. Directly, it implies in structural changes, investments, operational commitments in the scope of current routines, which are falling into obsolescence. In the Brazilian reality, this can also imply business possibilities, such as the creation of bundled services as well as an electricity offering based on seasonal or real time prices. This last topic will need a new organization on tariffs regulation.

In Brazil, although some critical components are already monitored, such as interface and audit meters as well as large customers meters, the data analyses are not systematic nor in real time for all dealers. In the best of cases, when some kind of accounting science is applied to this information, this knowledge becomes sectored and used to support a business segment (such as, loyalty of free (large) customers, those who can chose the structure of its supply, due to its high consumption).

In general, for the domestic customers in Brazil, there is no detailed information collected on the daily electricity consumption and their consequent analyses. Moreover, there is no specialization in the usage of such a piece of information and the knowledge gathered from it.

Broadening this monitoring and keeping the current Brazilian business model, with the registration of only monthly measurements, will incur in costs. For positive accountancy results, the large volume of information generated must be organized into a systematic, automated system, signaling the low-demand consumers information as a guidance of the usage (e.g., in order to detect theft, "leakage of electricity" or points to efficiency). This could derive and incentive strategic change in customer relationship, with a differentiated operational dynamic.

It may appear obvious that there will be a short-term return on the investment made in sensing (remote and tele-connected metering), mainly in regions where there is high default or electricity deviation. Although, this action must involve two subjective and relationship items: the commitment of the customer and value the electricity delivered services.

Incentives must be implemented to best practices and regulatory guidance, mainly for the regions or sub-regions with low-consumption customers and social commitment, seeking a

http://dx.doi.org/10.5772/62582

cultural change. Actions to make this consumption more efficient, and the understanding of the specific regional needs, may guarantee the breaking of the cycle regulation-cost-default-cut-theft. The creation of income conditions and the broadening of the feeling on electricity value, respecting both commitments and rights, are very important to minimize these issues that are both social and cultural in nature.

The analysis of Brazilian domestic consumption also brings the singularity of income distribution considering, that have an average monthly expenditure in electricity of less than 30 € (or less than 156 KWh) [9], with obvious implications in the business structures of the regional dealers.

Storage and retrieval information: these aspects are related to the legacy information technology systems, many times inappropriate to store and organize huge data volumes collected and exported in nearby real time. This problem, or from a pragmatic point of view, this solution, is a current structural practice in telecommunication companies, which have, historically, similar requirements to sense and supervise its network elements, its customers individually, as well as its entire capturing and data exchange system (boarder measuring, registration for tickets and clearing), geo-positioning of assets and interconnectivity, as well as price composing and billing according to timing and usage.

This cultural and structural change in storing and processing high volumes of data presents an appropriate cost-benefit relationship currently, with the technological evolution/availability of servers, storages, and cloud computing [10], ensuring strength and marketable products. Nowadays, in the Brazilian market, the electricity dealers are updating and also making investments in information technology, as well as changing its control processes. This is, therefore, a very favorable moment of (re)planning investments and organization for a differentiated operation, incorporating telecommunication infrastructure into the grid or assuming it as part of the new core business.

However, it is important to conduct an extensive cultural change work and improve operational processes for the future "smart business."

Information qualification and systems interoperability: these topics must be warned, according to smart grid current specifications, norms, and standards. Information capture and transfer (communication technologies), data gateway and applications need systems and suppliers interoperability. From the energy demand preview to increasing the share of renewable sources and taking advantage of intelligent network technologies and customer relationship, it is important to solve the present lack of knowledge and data necessary to manage the new business.

Basically, tests conducted with smart grid sensing focus on the qualification of the communication requirements, as well as the validation, quantification, and characterization of relevant parameters for an effective sensing, according to the perception of the dealer or energy company (standardization examples can be found at IEC Standards [11, 12]).

It is also important to mention, in a way that is coherent with its responsibility focused on the current energy service business, the lack knowledge and training in communications for

Brazilian electricity companies. It is worth mentioning also, the restrictions related to the Brazilian regulation model to electricity dealers to operate and offer different services from electricity, such as in the area of telecommunications. The obsolete or low networks connectivity is an action field. This knowledge is necessary for the maintenance of a sensing infrastructure and remote measuring requirements into a smart metering network. It is also both a business opportunity and a commitment. In the United States, this issue is also handled by the Federal Communications Commission (FCC), as mentioned in reference [13], reinforcing the involvement of several knowledge sectors for the composition of a business solution with all the necessary guidance.

In Brazil, this evolution and "joint" business opportunity among the electricity dealers and telecommunication system operators is a future commercial assumption, although some dealers have already ventured in studies on the supply of basic telecommunication services using their own infrastructure and the capillarity of their electricity grids.

Increase in system capacity: basically canalize efforts to build or reinforce the capacity in the high-voltage systems. The building of lines and transmission circuits must also characterize investments toward the (re) structuring of substations, adding criteria of robustness and failure tolerance, the broadening of control centers, systems, and protection and relay schemes. For instance, the great distances to be overcome to deliver energy generated in hydroelectric plants planned in Brazil, in the basin of Madeira River (Jirau with 3300 MW and Santo Antônio with 3150 MW), in the Amazon region, in the North of the country, to the cities in the Southeast, are engineering challenges yet to be solved.

There are approximately 2400 km of transmission lines to be built, with all their environmental impact still under studies. The interconnection, protection and operation of the system with this new generation are questions yet to be answered [14].

Coordination of areas, regions, and national control system and integration of electricity grids: this sector clearly must have special attention. A series of interrelated structural coordination roles must be conducted for an economic and trustworthy operation of the electrical system. Charging compensation and balancing, generation system coordination, transportation and distribution dealers, electricity market operations, government and emergency operation centers are included. The elements of smart grid in this context might include the collection of measurements of the entire system to determine its state and quality of electricity, and coordinate actions to increase the economic efficiency, reliability, environmental compliance, and respond to disruptions or systemic failures.

The need for regulation concerning these functions in the integration arena is evident, as well as the updating existing Brazilian control system. This system is presented as robust, but its conditions of self-adjustment, control, and recovery in case of simultaneous failures, of isolating problems and rebooting still need to evolve. The blackout events in several Brazilian regions from 2010 to 2014 had shown that the interconnected national system control must be improved and needs more intelligent and dynamic mechanisms to the decision making recovering procedures. These occurrences lighted the generation infrastructure and the composition of the electricity matrix as well as the transmission grid resources.

More and more actions and operational indicators must be defined, obtained, controlled, and managed by operators and by systems, a great challenge for the Interconnected National System [15].

System's bottlenecks and self-recovery control: controls for eliminating or at least recognition of the attention points or controlled overload. Together with the analysis of the system capacity, functionality includes increase power flow, enhanced voltage support, manage fault currents, allowing the operation, reaction and recovery from failures in the system in an effective and dynamic base. Surely, much technology is yet to be developed toward this effective control, such as power electronic devices, power electronic circuit breakers and others controllers, from the high-voltage grid control to the distribution grid.

The focus on robust interconnectivity, failure control, and recovery is evidenced, mainly when the aim is to ensure the automation of real time actions. It also reinforces the importance of regulating, guiding, and controlling entities on the interconnectivity between dealers, vendors, grids, generation fabrics, and systems. Many interoperability and multi-vendor tests have been made to ensure the robustness and self-recovery in the structures of the Brazilian transmission and distribution networks. Residential microgeneration are starting up into Brazilian energy grids and the two-way connection and control are in the initial evaluation.

Quality indicators: must be resultant from implementation and used to demonstrate systems efficacy. One polemic issue in the set of guidelines from the Brazilian regulatory agency (ANEEL) is associated to the models organization that figure out the delivered energy quality and the indicators reliability that show operational performance of the systems and their interfaces. These guidelines are presented in the models proposed for the reference company in the electric sector and in public referendum for the electronic meters [16, 17]. It is expected to reach the offering of services guided by levels (SLA—Service Level Agreement), such as in the telecommunication market.

The near horizon (until 2020) signals the exchange of 73.5 million meters in Brazil [9], in a migration to electronic metering technology, and if possible, intelligent. It will seek updating the installed metering devices and the entire measuring system, as well as improve electricity supply quality, reducing operational costs for distributors, fighting the losses and aiming toward energy efficiency [17].

The establishment of commitments is questioned as subject as implementation, tariffs, and incentives conditions which need to be granted that the costs of this process could be feasible in the current dealer regulation structure and to the customers. The vendors have been mobilized to supply devices and systems. Those should validate the requirements of interoperability, standardized interfaces, and certified by Brazilian measurement entities, following standards that are also under analysis.

Connectivity (broadened) empowerment for consumers and tariff model: all the prior functionalities are reflected on the end user care, recognized as the relationship on the customer point of view. This broadened view directly shows on the offering of connected services to the delivery of electricity (e.g., additional information for billing and real-time pricing (according to criteria established depending on the demand and load shape objectives), evaluation started by

ANEEL [18]), value-added services (such as safety and monitoring applications), and services involving the existing or added electricity infrastructure, established by smart grid implementation (such as Internet and data communication services).

It is intentional in the evolution of the Brazilian tariff model [18], several changes in the form of dividing the tariff components among the several users of the system. This shall cause specific tariff variations for each customer, depending on the group/sub-group/tariff category of the consumer, its consumption profile, as well as tariff flags creation and the dealer tariff review performance process. These flags must be extended to all the low-consumption customers (domestic and others), with signals in three time points: at the peak, intermediary, and out of the peak. Its implementation and viability are conditioned to the implementation of electronic meters (substitution of current electro-mechanic by electronic ones that allow the registration and differentiation of consumption by hours in a day). According to the agency, this change must not involve other expenses to consumers.

This discussion is also under regulation and must provide the conditions for the necessary evolution at the onset of structural changes for the intelligent grid.

4. Metering evolving and demand proving

According to Altvater [19], presented in [22]:

> The environment is not a limiting factor as long as it doesn't require too much with regards to the absorptive capacity of the global ecosystems. But a capitalist, industrial society is expansive in time and space; it grows, and rapidly. Even with zero growth, which is viewed by a number of ecologists as the solution for environmental problems, electricity and raw materials are consumed, despite zero or negative economic/monetary growth. It may even be that, with zero growth, the environmental burden becomes greater than with growth, because of the need to spare costs in the economic system. Therefore, the problem lies not in the dimension of the economic growth rates, but in how this "metabolism" is regulated: the material exchange between nature, individuals and society.

"Humans use natural resources (in the realm of the expanding economic system) progressively, as a source and as a deposit for undesired products." The advancement and incorporation of the environmental revolution's principles (energy efficiency, recycling, pollution control, and environmental design) should advance within the productive system through a sustainable guidance of public policies [20].

Therefore, in terms of energy efficiency and water demand management, these resources must be rationalized, but simply monitoring consumption is not enough. A cultural change in a population's habits must also be achieved

Public policies demand involvement and active citizenship by the consumers and require that the production-generation-delivery-consumption chain be monitored and audited. Organized market conditions are required from a consumption standpoint as are the amounts actually consumed and their financial counterpart (amounts paid). To achieve this goal, data (as information and as knowledge) must be organized, the amounts saved due in consumption habits must be indicated and a trend survey should be held (not only for the consumer but also for the utilities). It is also essential to monitor the entire system. Monitoring information allows incidents such as leaks to be detected and provides a review of operation procedures and consumption incentives and awareness to achieve the desired efficiency.

Digitalization and the advancement of the current control and telecommunications systems, as well as the lower costs of these systems, have allowed new operational possibilities. The current demand is for water and electricity management tools that will also allow customers to monitor and control their consumption in specific points of their residence. Some kind of system must be provided, as a platform for analyzing and monitoring consumption, providing automation of remote residential metering and sub-metering (internal to the residence). This platform must be integrated into a dedicated server and knowledge structure (centralizing the processing of meter readings), allowing behavior qualification, evaluation the actions taken for consumption reduction and/or awareness of use, as well as the planning of educational actions and grid expansion warnings.

In physical terms, the relationship environment receives the measurements collected from both commercial electronic electricity and water meters. Partial consumption measurements within households must be determined through water and electricity sub-meters (or smart plugs) installed at points of interest. Communication resources must be coupled to these devices so that the individual measurements can be transmitted to the centralized processing center at pre-set timeframes.

Additionally, this integration environment should be designed as a tool for Internet access and digital inclusion by supporting and valuing the communities' web knowledge and use. This feature naturally integrates into the global trends regarding service and universal access to information, access to knowledge, discretion over the use of services and serving the basic needs of the underprivileged populations.

"Management of natural resources may, in theory, through planning, anticipate, prevent and mitigate environmental impacts, as scientific knowledge allows this, and public pressure and demand make the policies become enforceable" [21]. However, there is a need for corporate and government policies to encourage cultural and educational changes related to conscious consuming.

With this platform, the energy industry should contribute with solutions that expand the availability of information and provide conditions for conscious, participatory decision making regarding consumption and thus promoting changes in habits and culture. It allows for a new perspective on the electricity and water businesses and in client/consumer relationship, with the latter auditing and verifying the efficiency of the former.

4.1. Field tests and results

Some clients' energy usage scenarios were prepared to validate and spotlight the challenges some company should find at smart grid/smart metering implementation. The researches look for governance and solutions for organizing and controlling information for customers within this new paradigm of knowing their actual consumption. One of that had the financial support of FINEP—Financiadora de Estudos e Projetos do Brasil (Study and Project Funding for Brazil) —from its 2008–2010 Economic Subsidy program. It resulted in a test lab and research on use, allowing the monitoring of the system, of the data collected and of the participant's level of involvement [22]. Another larger test was done by Inter-American Development Bank (IDB) support at one of the largest energy distributer company in Brazil as part of its smart grid field tests. Some results are presented below.

Two test fields were established at the initial lab tests [22]. The first lab consisted of a low-income housing complex—a vertical environment with several residential consumer units. The second test field was represented by a commercial consumer (horizontal environment), in an area of 2000 m^2, at an entity that deals with the autistic people education. They consume a lot of electricity and water and need to go through an energy efficiency and management better infrastructure. Management tools for efficient electricity and water were made available by treating the consumption measurements of each customer site. These measurements were collected through sensors and commercial electronic meters approved by the Brazilian metrological institute, composing an infrastructure designed for this purpose that integrates the electricity and water supply services, with the addition of telecommunications resources.

Another test field involving two hundred families (from different social classes at the same geographical region in Brazil) was organized to demonstrate a larger influence of efficiency improvement to those families, lighting the local habits and home appliances usage, as well as the recognition the real energy demand. Training, speeches, and monitoring actions were done to guarantee social compromising and conscious energy habits during the available pilot analysis.

Meters and home plug appliances used:

1. General meter: customers had installed meters that represent their monthly energy usage, with a timeframe of 1 information per minute;
2. Home appliance meter or logger: synchronized by a local router, at least 4 loggers per customer site were provided, with a timeframe of 1 information per minute available per logger and home device, organizing a reference warehouse to consumer habits analysis.[3]

First, the research demonstrates that it is necessary a sustainability, performance, and robustness of the technological solution for measuring consumption and qualifying its efficiency, integrating the electricity, water supply, and telecommunications services. It was necessary to integrate the consumer in this process, as a client and decision-maker, as agents in charge of their conduct that demonstrates the commitment to the planet and to their community.

From the consumer's standpoint, the improvement presented may broaden their individual opportunities as citizens through digital inclusion also associated with a more extensive

[3] One minute timeframe used at the pilots permit a near real-time energy information. The loggers and meters provided consumption data (Wh—watt hora) and peek of energy required (W). One can easily infer the nearby energy demand, by household, by circuit, and else by region. Weekly and monthly information provided important analysis of customer habits and derivate actions, from grid maintenance (to avoid grid or transformer overload and better quality of energy offered) to customers' efficiency speeches (focusing on how to avoid peak energy usage time, better home appliances performance and habits × comfort compromises).

culture that indicates their consumption; for example, it allows discretion in the use of the utilities. The pilot environments become a live laboratory, with the synergy of the technological elements and the community.

Smart meters and systems provided consumption history, daily consumption reports, and forecast savings for the undergoing month arising from changes in consumption habits. It allowed monitoring and controlling daily consumption pattern, as well as the monetary value that may be saved, depending on consumption habits and comfort decided. Additionally, it provided useful information for a life and resources governance commitment.

For the process governance, the impacts related to the sustainability of the energy demand were organized to represent some consumption indicators and features. The architecture organization is presented at the **Figure 1**. Some indicators are organized, according to [22], into some features as

- administrative functions to control user and customer management services and to control meta-data, events, and consumption ranges;
- ranking by consumer unit consumption;
- total consumption;
- consumption at peak times;
- percentage of total consumption (meters and sub-meters);
- indicators of consumption stationing level;
- indicators of drastic increase/decrease in the level of consumption;
- indicators of fluctuations in consumption levels (ping-pong effect);
- indicators of attempted intrusion or damage.

The functions listed below were defined to meet the consumers' needs:

- monthly consumption records;
- storage of images of the customer's premises;
- consumption profile;
- consumption in the past 12th months for each meter and sub-meter;
- monthly comparatives;
- comparison of the month's consumption with consumption averages;
- percentage per sub-meter;
- maximum, average, and minimum consumption;
- consumer unit comparison with ranges;
- consumption reduction tips based on performance;

- consumption forecasts for the current month and the economic expenditure forecast arising therefore;
- consumption targets and subsequent analysis;
- general tips for saving energy and water (leaflets).

Figure 1. A smart metering knowledge architecture organization.

The second customer field test was organized to evaluate the customer equipment usage (habits) as well as the energy demand. A consumer habits survey started the process, looking for customer energy efficiency compromising and home appliances usage recognition. Five hundred customers were pre-selected, two hundred customers participated in the survey and had smart metering and appliances home plugs installed. Based on their declared energy habits and home appliances usage an estimated load curve was done presented at **Figure 2**. For example, between 7:00 and 10 PM, energy peak usage represents families' habits, watching TV, washing clothes, and take their shower at the same time (using electric showers, a historical Brazilian way to locally heat shower water).

At the survey, the families declared their habits and the home appliances specifications that helps the pre-evaluation done. Others appliances were studied but electric shower, refrigerator, TV, and washing machine were more representative of the energy demand at the region under analysis.

http://dx.doi.org/10.5772/62582

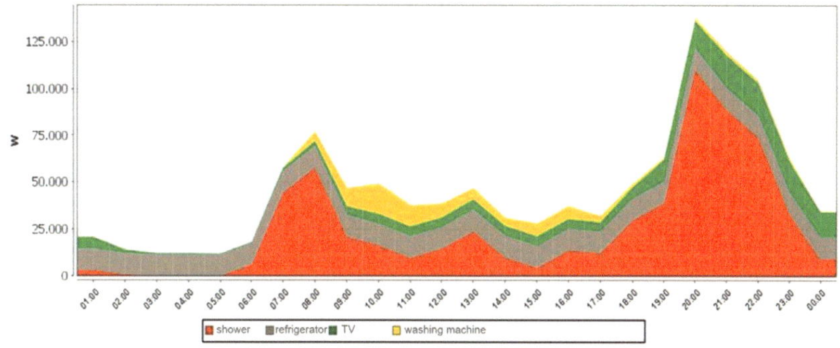

Figure 2. Estimated daily load curve from the major home appliances at households under analysis.

A daily demand (based on a 1-minute timeframe measures) proved the estimated load curve and is presented at **Figure 3**. This kind of information from the smart metering, energy usage day by day, helped the distribution utility at its grid evaluation and energy provisioning, as well as thief prevention, quality of energy qualification and regional services offerings. Smart metering as a tool could provide a business improvement, from grid maintenance to customer support organization.

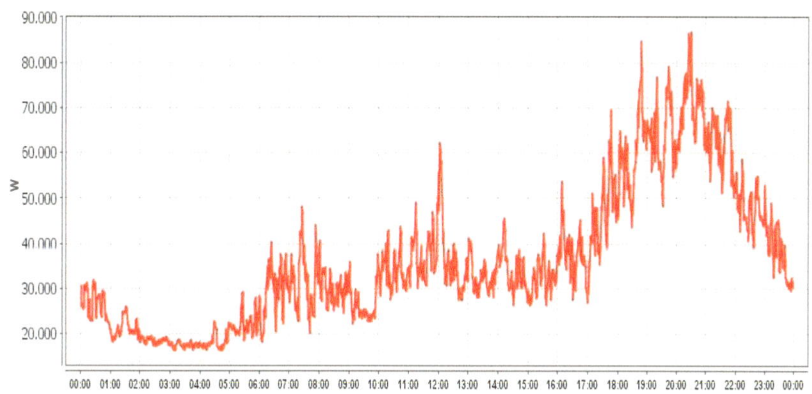

Figure 3. Daily demand measuring.

The historical energy usage (monthly usage—**Figure 4**) and daily demand (**Figure 5**) were presented to the customer, in a friendly way, involving the family with the energy efficiency compromising with their own energy usage performance. Mobile apps were provided to a daily home energy performance, with tips of better energy usage and costs simulation. After the test period, the families were conscious of their energy usage, appliances obsolescence as well as lighting influence, social and environmental compromising. After that, they could decide about their comfort and costs associated with. This research improved the energy distributor relationship and the knowledge about the regional expectation with smart grid (metering) implementation and possibilities.

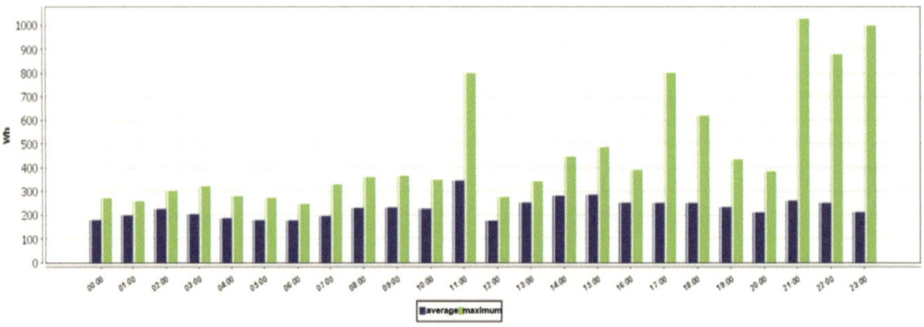

Figure 4. Customer's hourly average/maximum consumption during a month.

Figure 5. Seven days' customer's measuring (1-minute timeframe).

Basically, the analysis of the smart grit concept in Brazil is complex according to the effects of modernization and natural obsolescence on energy efficiency, the return on investments, communication and information safety, multiple suppliers, integration, and standardization. These considerations bring to light the commercial interests of meter vendors and concerns associated with demand increase and country energy planning.

In addition to metering issues, some other factors must be addressed to ensure effective grid operation as the integration of distributed generation, micro-generation, storage resources, and demand response infrastructure. The products that are used and will be used by customers, such as intelligent appliances and electric vehicles, are important components in this field of study, as well as generation of renewable energies derived from biomass, wind and solar sources.

With smart grid technologies, mainly smart metering, with continuous and dynamic data reading, demand information can be organized and used to improve performance and reliability of the electric system. For Brazil, these issues are also relevant, and there are presently, several ongoing experiments on distributed generation using technical approaches. Although, there has been great interest on smart grids and distributed generation, both within utilities and in academies, the extent directly related to consumer behavior is not explored nor their ability to modulate their demand, the answer to more efficient power use, reactions to fare variations, and even their interest in own generation investment opportunities.

Brazilian providers and regulatory agencies currently support research and innovation on the regulation of responsibilities and the requirements of future energy offerings, the incentives in the execution of projects, the implementation of advanced metering solutions, the evaluation of electricity quality, and the remote control of equipment. There are some experiments exploring Brazilian smart cities, including R&D projects as [23], looking for organizing and planning the new energy businesses.

The near future requires equating how this technological development and the role of consumers may affect the existing network, centralized generation system as well as deciding about their energy usage profile. It is necessary to consider more explicitly the role and behavior of the producer-consumer, without losing sight of the main objective of the network and generation system, which is to warrant reliable access, at lower cost, of power services for all users.

5. Conclusion

Sustainable development, within a context of social productivity and potential surplus to expand the frontiers of development, carries the contradictions and operational dynamics related to each social organization, and its ethics, culture, and history. Progress leads to questions concerning the advancement of individual, political, and economic liberties and of social opportunities within the context of relationships and the appropriate use of the environment to maintain diversities [24]. In Brazil, the uneven income and consumption distribution, migration, and urbanization lead to issues regarding sustainable development and direct government actions involving the investment in the rational use and quality of energy and water for the population, mainly for low-income communities. Given this context, education and tools should be used to show the individual efficiency in light of the collective well-being and the use of finite resources. Transparency and the participation of individuals and communities must also be included in the scope of audits and published [22].

Operating in this realm, the organization presented allows the management of energy efficiency in a sustainable manner, encouraging active citizenship by the users through the rational use of resources. Metering and friendly information could transform customer relationship, improve energy value recognition and social-environmental compromising, seeking to live up to the new management trends for electricity services and providing, as well, a fruitful field of information for the utility companies. This level of intelligence is possible when you take into account the sensory aspects and those related to information gathering by an advanced metering infrastructure (AMI) network, focusing on customers' services [25].

The progression of the networks, of the utility companies, of the consumers, of the regulation, and of energy use and generation use needs to be better articulated.

Incentives must be implemented to best practices and regulatory guidance, mainly for the regions or sub-regions with low-consumption customers and social commitment, seeking a cultural change, as well as control high default or electricity deviation. Actions to make this consumption more efficient, and the understanding of the specific regional needs, may guarantee the breaking of the historic cycle regulation-cost-default-cut-theft. The creation of income conditions and the broadening of the feeling on electricity value, respecting commitments and rights, are very important to minimize the issues that are both social and cultural in nature.

To sum up, a level of commitment to an energy efficient and sustainable model will depend on a number of rearrangements at the Brazilian energy industry and country governmental guidelines, such as the adequacy structure of generation and delivery, regulatory and standardization reinforcement, and also the understanding of customers' needs and socio-cultural efforts to motivate the conscious use of energy.

In consonance with the theme, the relationship with the regional client is presented as the basis for regulating service supply. Thus, new regulations involving the electrical and telecommunication industries must develop an anchor role and be used as guidelines for the developing of smart grid (and reinforce smart metering characteristics) in economic and industrial relationships guiding the development of this sector in the country.

Acknowledgements

Parts of Section 4 of this chapter are reproduced from author's earlier proceedings paper [22] from Energy (volume 45, Issue 1, September 2012—pages 528–540, © 2012 Elsevier)

Author details

Carlos Alberto Fróes Lima

Address all correspondence to: froes@knbs.com.br

Unicamp, University of Campinas, SP, Brazil

References

[1] Gellings, C. W. The Smart Grid – Enabling Energy Efficiency and Demand Response. USA: The Fairmont Press Inc., 2009, pp. 300.

[2] COP 16. 2010 United Nations Climate Change Conference. Available at: http://unfccc.int/meetings/cop_16/items/5571.php [accessed 01.02.2011].

[3] Electric Power Research Institute (EPRI). The Green Grid – Energy Savings and Carbon Emissions Reductions Enabled by a Smart Grid. Report 1016905. (2008)., 64 p. Available at: http://my.epri.com/portal/server.pt?space=CommunityPage&cached=true&paren-

tname=ObjMgr&parentid=2&control=SetCommunity&CommunityID=405 [accessed 15.01.2010].

[4] Sioshansi, F. P. Competitive Electricity Markets: Design, Implementation, Performance, in Plastics. Oxford: Elsevier, 1st ed., 2008, pp. 582.

[5] Goto, M., Yajima, M. A New Stage in Electricity Liberalization in Japan: Issues and Expectations, in Sioshansi, F. and Pfaffenberger, W. (eds.), Electricity Market Reform: An International Perspective, Oxford:Elsevier, 2006, pp. 617–644.

[6] Ministry of Economy, Trade and Industry (METI). New National Energy Strategy – Formulated by the Japanese Government – Japan Gas Association, Tokyo: online May, 2006 - Maio, 2006, p. 39. Available at: http://www.gas.or.jp/en/newsletter/images/05/pdf/newstrategy.pdf [accessed 23.09.2012].

[7] US Department of Energy (DOE). The Future of the Grid: Evolving to Meet America's Needs, 2014. Available at: http://energy.gov/oe/services/technology-development/smart-grid [accessed 16.01.2016].

[8] European Technology Platform (ETP). National and Regional Smart Grids Initiatives in Europe. Available at: http://www.smartgrids.eu/ETP_Documents [accessed 16.01.2016].

[9] ANEEL Database. Brazilian Consumption and Tariffs Information. Available at: http://www.aneel.gov.br/area.cfm?idArea=550 [accessed 26.01.2016].

[10] Srikantaiah, S., et al., Energy Aware Consolidation for Cloud Computing. Hotpower'08 Proceedings of the 2008 Conference On Power Aware Computing And Systems, Usenix Association, Berkeley, CA, USA, 2008.

[11] IEC 61850 - Communication networks and systems in substations, IEC Standard, 14 parts, 2002–2004.

[12] International Electrotechnical Commission (IEC). IEC Global Standards for Smart Grids. Available at: http://www.iec.ch/smartgrid/ [accessed 15.01.2016].

[13] DOE. Smart Grid Research & Development – Multi-Year Program Plan (MYPP) (2010–2014).. Available at: http://energy.gov/sites/prod/files/SG_MYPP_2012%20Update.pdf [accessed 10.09.2012].

[14] EPE (Brazilian Energy Research Company). Transmission studies of the Madeira River plants. Available at: http://www.epe.gov.br/Transmissao/Paginas/LeilaoMadeira07_12.aspx [accessed 04.03.2015].

[15] ONS (Brazilian National Electric System Operator). Energy Operation Planning. Available at: http://www.ons.org.br/procedimentos/index.aspx [accessed 23.08.2015].

[16] ANEEL (Brazilian National Electric Energy Agency). Reference enterprise. (2008). Available at: http://www.aneel.gov.br/aplicacoes/audiencia/dspListaDetalhe.cfm?

attAnoAud=2007&attIdeFasAud=266&id_area=13&attAnoFasAud=2008 [accessed 01.03.2011].

[17] ANEEL (Brazilian National Electric Energy Agency). Public consultation for subsidies and information for the implementation of electronic measurement at low voltage - CP 043/2010 and contributions. Available at: http://www.aneel.gov.br/aplicacoes/audiencia/dspListaDetalhe.cfm?attAnoAud=2010&attIdeFasAud=435&id_area=13&attAnoFasAud=2010 [accessed 20.02.2011].

[18] ANEEL (Brazilian National Electric Energy Agency). Public Hearing on new tariff structure, 120/2010. Available at: http://www.aneel.gov.br/aplicacoes/audiencia/dspListaDetalhe.cfm?attAnoAud=2010&attIdeFasAud=513&id_area=13&attAnoFasAud=2011 [accessed 20/02/2011].

[19] Altvater, E. Introduction: Why the development is contrary to the environment? The Price of Richness. São Paulo: UNESP, 1995, pp. 11–41.

[20] Viola, E. The Globalization of Environmental Policy in Brazil, 1990-1998. XXI International Congress of the Latin American Studies Association, Panel ENV 24, Social and Environmental Change in the Brazilian Amazon; The Palmer House Hilton Hotel, Chicago, 24–26 September 1998.

[21] Vilani, R., Machado, C. J. S. Energy and Environment in the Growth Acceleration Program (PAC): A Critical Analysis. IV Encontro Nacional da ANPPAS, GT: Energia e Ambiente, Brasília, DF, 4–6 June 2008.

[22] Froes Lima, C. A., Navas, J. R. P. Smart metering and systems to support a conscious use of water and electricity, Energy, Volume 45, Issue 23, September 2012, Pages 528–540.

[23] CEMIG. Cemig to Launch Smart Grid Project in Sete Lagoas, 2010. Available at: http://www.metering.com/cemig-to-launch-smart-grid-project-in-sete-lagoas/ [accessed 01.09.2013].

[24] Fróes Lima, C. A., Silva, A. L. R. A smart consumer requires intelligent services, requires intelligent service, and decides a smart grid, VI CIERTEC, Belo Horizonte, 2009, 10 p.

[25] Sui, H., et al., An AMI System for the Deregulated Electricity Markets, Industry Applications, IEEE Transaction, Volume 45, Issue 6, November 2009, pp. 2104–2108.

Section 2

Smart Metering Architecture, Implementation and Application

Chapter 3

Advanced Metering Infrastructure Based on Smart Meters in Smart Grid

Trong Nghia Le, Wen-Long Chin,
Dang Khoa Truong and Tran Hiep Nguyen

Additional information is available at the end of the chapter

http://dx.doi.org/10.5772/63631

Abstract

Due to lack of situational awareness, automated analysis, poor visibility, and mechanical switches, today's electric power grid has been aging and ill-suited to the demand for electricity, which has gradually increased, in the twenty-first century. Besides, the global climate change and the greenhouse gas emissions on the Earth caused by the electricity industries, the growing population, one-way communication, equipment failures, energy storage problems, the capacity limitations of electricity generation, decrease in fossil fuels, and resilience problems put more stress on the existing power grid. Consequently, the smart grid (SG) has emerged to address these challenges. To realize the SG, an advanced metering infrastructure (AMI) based on smart meters is the most important key.

Keywords: advanced metering infrastructure, communications, security, smart grid, smart meters

1. Introduction

An electric power grid is a network of power generators, transmission lines, transformers, and distribution/relay systems to provide its consumers (residential, industrial, and commercial) with the power they need. Currently, electrical energy is generated in centralized power plants and transported over a long-distance transmission network to distribution networks before reaching the end consumers via communication and power flows in only one direction, i.e., from power plants to the customers, which is collectively called an electric grid. After many decades of development, it has been realized that various utilities can interconnect to achieve greater reliability of overall power system by compensating for unexpected failures as well as disconnections from power devices, i.e., transmission lines and generators.

In an electric grid, generation, transmission, and distribution of power should be precisely coordinated. **Figure 1** depicts various sections in today's electric grid, which consists of four areas that are generation, transmission, distribution, and customers [1]. Generation involves the production of electricity from energy sources such as wind and solar farms, coal plants, and hydroelectric dams. Because generators cannot be located too close to population centers for safety, legal, and financial reasons, the electric grid needs transmission lines to carry the electricity over long distances (often more than hundreds of miles). Distribution includes taking the electricity from transmission lines and delivering it to the customers. Typically, an electricity distribution system includes medium voltage power lines (below 50 kV), substations, and transformers, starting at the transmission substations and ending at the meters of customers. A substation consists of a bus to split up the power into different regions, step-down transformers, relays, and circuit breakers, which are designed to disconnect the substation from different distribution lines or from the power grid when necessary. The same transmission substation can deliver the power at different voltages to different regions, and the power might be further stepped down in several stages to reach 7200 V. A transformer is used to reduce voltage from 7200 to 240 V at each customer site. Two wires from the trans-

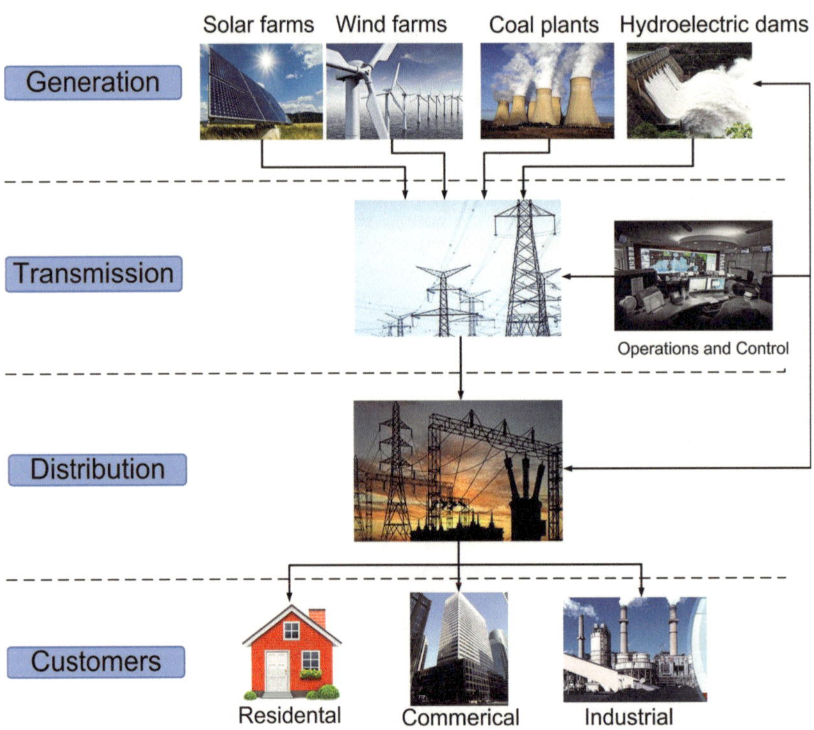

Figure 1. Typical electric power grid [1].

former are used to connect to power meters at a building or house, each carrying 120 V. These two wires are 180° out of phase, resulting in 240 V, which allows customers to use both 240 and 120 V appliances.

Due to the lack of situational awareness and automated analysis, today's electric power grid has been aging and ill-suited to the fast growing demand for electricity in the twenty-first century [2]. For example, in the United States, the consumption and demand for electricity have increased by 2.5% annually over the past 20 years [3]. Besides, the global climate change and greenhouse gas emissions on the Earth caused by the electricity and transportation industries [4] put more stress on the existing power grids. Consequently, a new concept of next-generation electric power system is urgently needed to address these challenges, which motivates the proposal of smart grid (SG).

The SG can be viewed as a superposition of communication networks on the electric grids. Hence, it can improve efficiency, reliability, safety, and security of electricity supply to the customers, with a seamless integration of renewable and alternative energy sources, such as photovoltaic systems, wind energy, biomass power generation, tidal power, small hydropower plants, and plug-in hybrid electric vehicles, through automated control and modern communications technologies [5]. In SG, various components in these four areas of the electric grid are linked together via two-way communications and power flows to provide interoperability among them. Thus, consumers can not only draw power but also supply surplus power to the grid using smart meters that enable monitoring and measuring of these bidirectional flows. This new infrastructure could potentially produce millions of alternate micro-energy sources and allow improved load balancing through instantaneous electricity demand information exchanges, which could help power plants match their output to demand with the help of information generated from metering, sensing, and monitoring.

To realize the SG, an advanced metering infrastructure (AMI) based on smart meters is the most important key. The AMI is the system that collects and analyzes data from smart meters using two-way communications, and giving intelligent management of various power-related applications and services based on that data. The AMI is the deployment of a metering solution with two-way communications to the electric meter. The implementation of AMI is widely seen as the first step in the digitalization of the electric grid control systems. Recently, AMI has gained great attraction in both industry and commerce due to the accurate improvement in online meter reading and control. The AMI is the architecture for automated two-way communications between smart utility meters and utility companies. The AMI includes smart meters, e.g., electric, gas, and heat meters, at customer premises, access points, communication backbone network between customer and service providers, and data management systems to measure, collect, manage, and analyze the data for further processing. The smart meter can identify power consumption in much more detail than a conventional meter and periodically send the collected information back to the utility company for load monitoring and billing purposes. Besides, the data from smart meter readings are also critical for the control center to implement Demand/Response mechanism. By using smart meters, customers can control their power consumption and manage how much power they are using, particularly managing the peak load. Hence, through customer participation, the utility companies can likely provide electricity at lower and even rates for all their customers, and the consequent carbon dioxide emission will decrease. Despite the increase in the utilization of AMI, there has been very little assessment or research and development effort to identify the security needs for such systems. Hence, the aim of this chapter is to offer a comprehensive description about AMI based on smart meters in SG. In addition, the issues on security, major challenges, and solutions in AMI in SG are also proposed.

2. Smart meter architecture

Smart meter is an advanced energy meter that supports two-way communications compared with a conventional energy meter. Hence, it can measure the energy consumption data of a consumer and then transmits added information to the utility companies to support decentralized generation sources and energy storage devices, and bill the customer accordingly. Besides, smart meters can receive information about electricity price and commands from utility companies and then deliver them to consumers. In practice, smart meters can read energy consumption information of customers in real time, such as values of voltage, frequency, and phase angle, and then they securely communicate the information to control centers. By using bidirectional communication of data, smart meters can collect information regarding the electricity consumption values of customer premises. Data collected by smart meters is a combination of parameters such as a unique meter identifier, timestamp of the data, and the electricity consumption values. Based on the information, smart meters can monitor and execute control commands for all home devices and appliances at the customer's premises remotely as well as locally. Besides, smart meters can communicate with other meters in their reach using home area network (HAN) to collect diagnostic information about appliances at the customer as well as the distribution grid. Moreover, smart meters can be programmed such that, only power consumed from the utility grid is billed whereas the power consumed from

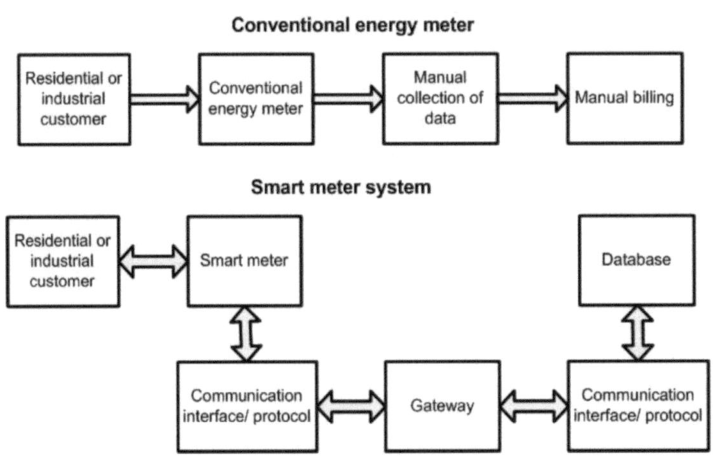

Figure 2. Architectural model of conventional energy meter and smart meter.

http://dx.doi.org/10.5772/63631

the distributed generation sources or storage devices owned by the customers is not billed. As a result, they can limit the maximum electricity consumption, and can terminate or reconnect electricity supply to any customer remotely [6]. **Figure 2** shows an architectural model of a conventional energy meter and a smart meter.

A smart meter system includes various control devices and sensors to identify parameters and situations in SG and then it transfers the collected data to the control center or provides command signals to the devices in the home of customers. The collected electricity consumption data from all devices of customers on a regular basis helps the utility companies to manage electricity demand/response more efficiently and also to provide useful information to the customers about the cost-efficient methods to use their appliances. Besides, smart meters can be programmed to maintain a schedule for operation of the home devices and control operation of other appliances accordingly, i.e., to control light, heat up water in swimming pool, air conditioning, washing machine, and other appliances [7]. In addition, by integrating smart meters in electricity grid, utility companies can detect and identify electricity theft and unauthorized consumption in view of improving the power quality and distribution efficiency [8]. Hence, smart meters would play an extremely important role in monitoring the performance and the energy usage characteristics of the load on the electricity distribution grid in the future.

Typically, smart meters implement two major functions, which are communication and measurement [9]. Hence, each meter is equipped with two subsystems as communication and metrology, respectively. The communication part includes security and encryption that define the suitable data transmission approach. The metrology varies depending on multiple characters such as measured phenomenon, technical requirements, region, accuracy, applications, and level of data security. Regardless of the type or quantity of their measurement, smart meters should have six basic functionalities as mentioned [10], which include the following:

Quantitative measurement: Smart meters have to accurately measure the quantity of the medium by using various topologies, physical principles, and approaches.

Control and calibration: Smart meters should be providing ability to compensate the small variations according to each system type.

Security communication: The meters have ability receiving operational commands and sending stored data as well as upgrades for its firmware trustworthily.

Power management: Smart meters have to help the system to exactly maintain its functionality when the primary source of energy is lost.

Display: Smart meters will send and display information usage of electricity energy to customers for billing in real time. Besides, the information of real time consumption displayed on smart meters helps customers to manage their demand efficiently.

Synchronization: Typically, smart meters transmit data of customers to the collector systems or central hubs for billing and data analysis. Hence, timing synchronization is very important for reliable transmission of data, particularly in case of wireless communication.

As a result, based on smart meters, utility companies can provide highly reliable, readily accessible, flexible, and cost-effective energy services to their consumers by combining advantages of both small distributed power generators and large centralized generators. Moreover, demand side management techniques require that these companies have to collect large quantity of data from smart meters in real time. One of key components to implement this concept is advanced metering infrastructure, which collects and analyzes data from smart meters, and gives intelligent management of various power-related applications and services based on that data. In next section, we present AMI based on smart meters.

3. AMI based on smart meters in SG

3.1. AMI architecture

AMI is a main mechanism for the realization of other smart grid applications to deliver operational and business benefits across the utility. AMI is the system that collects and analyzes data from smart meters using two-way communications between user domain and utility domain, and gives intelligent management of various power-related applications and services based on that data. The implementation of AMI is widely seen as the first step in the digitalization of the electric grid control systems. AMI's main functionalities encompass power measurement facilities, assisting adaptive power pricing and demand side management, providing self-healing ability, and interfaces for other systems. Recently, AMI has gained great attraction in both industry and academia due to the accurate improvement in online meter reading and control. AMI helps for financial benefits, improved services, and opportunities for consideration of environmental concerns.

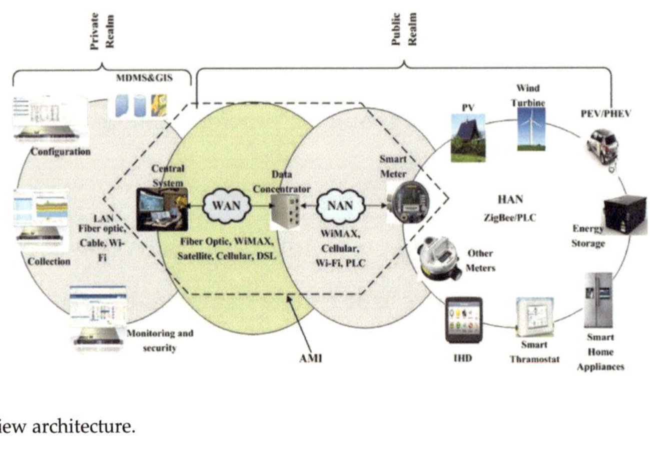

Figure 3. AMI overview architecture.

The AMI includes smart meters, e.g., electric, gas, and heat meters, at customer premises, access points, communication backbone networks between customers and service providers, and data management systems to measure, collect, manage, and analyze the data for further processes. These AMI components are usually located in various networks [11] and different realms such as public and private ones [12]. In AMI systems, smart meters are regarded as the key interfaces for physical, information, and social domains of the smart grid. **Figure 3** shows AMI overview architecture that is integrated in a broader context of power generation, transmission, distribution, and customer using HAN, neighborhood area network (NAN), and wide area network (WAN).

From this figure, we can see that the smart meter is a key device for consumers because it is responsible for monitoring and recording power consumption of home appliances. HAN provides connections between home appliances, other integrated systems such as rooftop photovoltaic system, distributed sensors, plug-in electric vehicle/plug-in hybrid electric vehicle, in-home display (IHD), smart thermostat, etc., and the smart meter. For communicating among these constituents, power line communications (PLCs) or wireless communications, such as ZigBee, 6LowPAN, Z-wave, and others can be utilized. NAN provides communication links between a number of individual smart meters and a data concentrator using WiMAX or cellular technologies. A number of data concentrators are connected to a central system (it also is called an AMI headend) in the utility side through WAN. Typically, WAN consists of two interconnected networks, i.e., the core networks and backhaul networks. The core networks provide connections to the control center and commonly use fiber optics or cellular networks to guarantee high data rates and low latency. The backhaul networks handle the broadband connections to NANs and monitoring devices. Applying cognitive radio (CR) technology in backhaul networks contributes to reducing the cost for investment and enhancing the flexibility, capacity, and coverage. Typically, the AMI headend, which is located in the utility side, includes geographic information system (GIS), configuration system, meter data management system (MDMS), etc. These subsystems can utilize a local area network (LAN) for intercommunication. In the next section, we present detail in AMI communications infrastructure.

3.2. AMI communications infrastructure

In AMI, the smart meter can identify power consumption in much more detail than a conventional meter and periodically send the collected information back to the utility company for load monitoring and billing purposes. In addition, the data from smart meter readings are also critical for the control center to implement demand response mechanisms. Using smart meters, customers can control their power consumption and manage how much power they are using, particularly managing the peak load. Hence, through customer participation, the utility companies can likely provide electricity at a lower rate for all their customers, and the consequent carbon dioxide emission will be decreased. Typically, existing AMIs collect data from smart meters and sensors with intervals of 15 min, the collected data are huge and important, and it is estimated that a moderately sized city with 2 million homes could generate 22 GB of meter data every day [13], and is referred to as "Big Data," easily overwhelming best planned data center capacity in a fairly short time. In particularly, MDMS with the analytical tools is considered the central module of the management system. Besides, MDMS has to ensure complete and accurate Big Data from customer to the management modules under possible interruptions at lower layers by performing validation, estimation and editing on the AMI data. Moreover, the distribution network automation system, which collects up to 30 samples per second per sensor for real-time control of SG [14], third-party systems, such as storages or distributed energy resources, connected to the grid, and asset management system responsible for communication among central command are also sources created Big Data in SG. As a result, the communication backbone networks should be reliable, secure, scalable, and cost-effective enough to meet the requirements in terms of bandwidth and latency to communicate the data.

In [15], by deploying an AMI, reliability, operational efficiency, and customer satisfaction can be achieved. This chapter also suggested several additional benefits gained in the AMI, such as managing power quality and asset management to improve service of the utility company. However, a robust communication backbone for the AMI data transmissions was not provided in the chapter. In particular, the AMI communication models include thousands of smart meters, multiple access points, and a mesh network, which is formed between smart meters for data routing purposes using industrial, scientific, and medical (ISM) frequency bands. Meanwhile, the aggregated data are routed to the utility company by access points mostly using licensed bands. The reliability and security of data communications between AMI components suffer from crowded and noisy ISM bands in urban areas. Packet losses, performance degradation, latency, and signal interferences are some of the consequences of heterogeneous spectrum characteristics of the crowded wireless communications. Moreover, the use of licensed bands to communicate the data between access points and utility companies requiring extra costs, which is another obstacle to deploy AMI in SG. Consequently, providing a robust communication backbone is sometimes hardly achievable, and it also comes with some obstacles for implementation of AMI in SG.

Several works investigated integrated communication technologies for the communication backbone of AMI. For example, mesh, Ethernet, and cellular AMI network topologies for SG have been proposed in [16–18]. In [16], the authors proposed mesh networks with a transmission architecture based on ZigBee, because the ZigBee protocol was integrated into smart meters by many AMI vendors, such as Itron, Elster, and Landis Gyr. The operation of ZigBee under an unlicensed spectrum makes it easy to implement network, being a standardized protocol based on the IEEE 802.15.4 standard. Nevertheless, ZigBee also has its own disadvantages, i.e., the transmission distance is limited, the rate of data transmission is low, and the capability to penetrate the barriers is weak due to non-line-of-sight transmission. Moreover, ZigBee may cause interferences to other appliances, which operate in the identical 2.4 GHz ISM frequency band, such as IEEE 802.11 wireless local area networks (WLANs), WiFi, Bluetooth, and Microwave. Inefficiencies of AMI based on ZigBee will arise when transmission distances increase. High levels of internetwork coordination are necessary in the deployment of new mesh networks. Improved alternatives of AMI mesh networks utilize IEEE 802.11 (a, b, g, n) protocols. However, such networks only support transmission distances ranging from 50 to 200 m, which is also problematic for robust metropolitan area coverage. To increase the transmission distances in the metropolitan areas and security of data communications between AMI components, in [17], the authors discussed communication infrastructure based on

Ethernet. The proposed method can support automated meter readings, customer home appliance connections, distribution automation, and substation automation. However, AMI based on Ethernet is not always affordable. In addition, the wireline systems can be challenging to rapid redeployment, particularly in emergency situations. To overcome this problem, the authors in [18] proposed a framework for Radio frequency (RF) mesh networking interfaced with high-speed access networks such as WiMAX. In the framework, the AMI smart meters are capable of two-way communications over a 900 MHz wireless mesh network back to a collection point at the substation. A private high-speed access network, which typically can be fiber or an existing cellular network such as WiMAX, will be then utilized to connect the substation to the corporate network. However, the AMI network topology based on cellular network or fiber for SG brings in extra costs to the utility companies and customers. Especially, the AMI interfaces for future proprietary protocols were not proposed in the framework. Ideally, AMI interfaces should be upgraded via software without hardware modifications to save time and cost.

4. Security in AMI

AMI security is required to protect both communication networks and power grid, because these two systems need to ensure their availability of access as well as survivability in different scenarios. However, the security of communication networks and power grid differ in several ways. In a communication network, latency needs to be limited and bandwidth needs to be guaranteed, whereas data manipulation (placement of false data), destruction of data, and unauthorized access should be prevented. On the other hand, security of a power grid needs to ensure reliability, power quality, and stability. Despite these differences, security between the two systems must be coordinated because the power grid and communication network can be used to launch attacks against each other. For instance, because the power supply in SG will be controlled by instantaneous users, information, manipulation of usage data could create a fictitious grid imbalance leading to voltage variations that can create large-scale failures. Similarly, if the state information of the grid is poisoned, the grid could be destabilized with a potential for physical damage. Physical damage could occur through overheating of transformers and relays or through voltage fluctuations in appliances. Due to the critical role of AMI in the SG, AMI security is special importance for the security of the SG. Given importance in AMI security, in [19] the authors discuss the security issue from two major aspects: maintaining the privacy of consumer's information and resilience of system against cyber or external attacks. Besides, the authors in [20] propose security in AMI using key management scheme for communication system. We can summarize these aspects as follows.

4.1. Privacy of end user's information

In AMI, smart meters are capable of collecting information of customers in every 15 min. However, current technologies even allow for collecting the data with intervals of minute [21]. Hence, if attackers analyze the data, they can achieve "consumer profiling" with an alarmingly high accuracy, for example, they know how many people live in the house, type of devices, duration of occupancy, ability of security and alarming systems. The profiling allows the attackers to extract behavior of customers without the need of using computer-aided tools or sophisticated algorithms. The authors in [21] have shown that they can identify the use of major devices in a house of customer by analyzing cumulative energy consumption data from the smart meter with a 15 min interval. Molina-Markham et al. [22] use the current general statistical schemes to identify the usage pattern from AMI data, which is valuable to third parties, such as entertainment agencies, insurance companies, and government authorities. In AMI of SG, you can access the network data using your name and address information collected and stored for billing purpose. From the detailed information, the process can backfire if it is used without your consent.

To discuss the importance of privacy, it is necessary to consider electrical behavior of an appliance while it is operating, which is defined as load signature (LS) because each appliance has different measurable behaviors. For example, consumption behavior of each electrical appliance is a signature, which could be measured at meter point. Typical variables are current, voltage, and power or energy. To protect the customers' privacy, a common method is to make it impossible for unauthorized parties to differentiate between load patterns and signatures. The authors in [23] proposed "load signature moderation" technique to facilitate customers' privacy protection by reshaping the overall pattern of data to make differentiating between load patterns and signatures impossible. This technique combines three methods, which are smoothing, hiding, and mystifying consumption, utilizing cooperation of grid and storage/battery as power source. The method is also defined as "undetectability" in [24].

4.2. Security against external cyber or physical attacks

The AMI-Sec Task Force, which is formed by security domain experts, industry leaders and standards bodies, developed the requirements for AMI security [25]. It provides guidance and security controls to organizations developing or implementing AMI solutions. According to the report in [25], security requirements for AMI system include confidentiality, integrity, and availability (or resilience to DoS attacks). Hence, the security for AMI system should satisfy the requirements as follows:

- **Confidentiality:** In the AMI, the metrology and consumption information communications have to meet confidentiality requirements to protect the customer's privacy and business information. This means physical theft of smart meters to access the stored information, unauthorized access to the data, as well as customer's access to other customers' data should be prevented. At AMI headend, only authorized systems are permitted to access to specific customer's information.
- **Integrity:** The AMI system should ensure the integrity of transmitted messages, as the operation of the AMI is dependent on the integrity of communicated information. Integrity in AMI means that the transmitted data from the meter to the utility as well as control commands from the utility to the meter and the received data from smart meters are maintained and protected from any changes such as malicious modification, insertion, deletion, or replay. The integrity of data can be ensured by using cryptographic techniques

to prevent hackers from pretending they are authorized entities and issue commands to perform their attacks. In AMI, smart meters have to detect cyber-attacks and ignore all issued control commands from attacker to protect the integrity of SG.

- **Availability:** The assurance that any network resources, such as data, bandwidth, and equipment, will always be available to any authorized entity. One of the important functions of the availability is to prevent denial-of-service (DoS) attacks, energy starvation, and selfishness. Hence, AMI components should protect against or limit the effects of DoS attacks. The AMI system should restrict the ability of internal or external users to launch DoS attacks against other AMI components or networks. Besides, the main reason for unavailability of data is component failure, such as communication failure (due to interference, cut cables, path degeneration, loss of bandwidth, network traffic, etc.), software problem, physical damage, or human tampering with the meter.

- **Accountability:** Also known as nonrepudiation or nondenial. Accountability techniques prevent either receiver or sender from denying a message by ensuring that undeniable proof will exist to verify the truthfulness of any claim of an entity. Accountability is specifically important for billing purpose as well as responses to control signals and in the actual metrology data. In AMI, the accountability requirement is a major concern, because different devices are usually owned by different entities, for example, service providers, customers, and they are manufactured by different vendors. To ensure accountability, time synchronization across AMI network as well as accurate time stamp of collected data is vital.

Based on the security requirements for AMI system was mentioned, security in AMI is very complex. Hence, just a single solution is insufficient for securing AMI. The authors in [22] present the threats to the security in AMI and then they propose some technologies as well as policies to improve the system's security.

4.3. Security in AMI using key management scheme

A typical AMI involves smart meters, HAN, NAN, WAN, and MDMS. For secure communication between these entities, confidentiality, integrity, and authentication should be guaranteed in the first place. Meanwhile, availability is also a critical requirement that should be fulfilled due to the high availability of electrical power. Besides, the AMI system must implement intelligent applications, such as dynamic electricity pricing, demand response, and real-time measuring/monitoring. Hence, AMI should be able to support different communication types (i.e., unicast, multicast and broadcast communications) for both customers and the utility companies to propagate information between the utility and smart meters [26]. Measured data are usually unicast communication from smart meters to the utility companies. Meanwhile, electricity pricing information is communicated multicast or broadcast from the utility to smart meters. Demand response program information is transmitted broadcast to all customers. As a result, by using the key management scheme for the AMI system, unicast, multicast, and broadcast communications should be able to securely and efficiently deliver [20].

To meet the security requirements for AMI, an underlying key management scheme is needed to generate and update keys for secure message transmission and authentication. Unfortunately, existing key management schemes designed for IT systems are simply inapplicable for AMI infrastructure in SG due to the reasons as follows:

- AMI is a complex heterogeneous system, which includes various entities with different computing ability, storage, and communication capability. In AMI, the smart meters are typical resource-constrained appliances, which have limited computation and storage capability. Meanwhile, the MDMS has high computing ability and plenty of storage resources. Hence, AMI utilizes the key management scheme, which not only achieves the security requirements of the system, but also accommodates this imbalance in its existing resources.

- Typically, AMI in SG is built based on combining IT systems with electric power system. Thus, problems of AMI are unique that are not encountered in traditional electric power system as well as IT systems. For example, electric power service demands the high availability, which is the same high availability of the security schemes in IT systems. The availability of electric power service and IT systems is considered as DoS attacks. As a result, the key management scheme must been designed with mechanisms to protect against DoS attacks. Additionally, the key management scheme has ability to support various modes of data transmission used in AMI.

- Because AMI may consist of a huge number of smart meters. Hence, the key management protocol has to open with scalable ability for such a big system.

Currently, in [26, 27], the authors propose key management schemes in AMI for SG. However, these schemes cannot completely satisfy the above requirements of security. For example, the authors in [26] present a new key management scheme for AMI, but this method is vulnerable to DoS attacks and inefficient in key management for a big system. In [27], the authors propose the key management scheme using physically unclonable functions to guarantee the security requirements of the system; however, the method is designed without open protocol with scalable ability for the big size of AMI. To overcome these problems, a hybrid key management scheme for AMI is proposed in [20] by integrating public key cryptosystem with symmetric cryptosystem. In this hybrid scheme, the elliptic curve cryptosystems are utilized to achieve efficient session key generation and trusted authentication. Besides, to generate and update group keys efficiently, the authors employ a specially designed key hierarchy.

Based on the security requirements of AMI, the system structure, and required availability, a key security technology using trusted computing methodologies and public key infrastructure (PKI) is proposed in [28]. By combining PKI technologies with trusted computing elements, the method is the most desirable solution for SG security as well as AMI. However, the method is complex, especially in the big system. To reduce the complexity of the method, the authors propose a technology utilizing the four major technical elements, namely automated trust anchor security, PKI standards, SG PKI tools, and certificate attributes. In [29], the authors complement a novel technical element to reduce the complexity of PKI security, which is device attestation. The proposed method includes the PKI elements into the overall security architecture to achieve a cost-effective and comprehensive solution for AMI security in SG. Besides, the trusted computing elements are utilized to guarantee that a malware cannot to access to the software processing devices. The main functionality of trusted computing is to allow any devices, which want to join a grid network, to verify that authorized code runs on that system. The adoption of strict code signing standards by SG suppliers and operators was also suggested in [28]. Mechanisms for enforcing such standards have been put forward by the Trusted Computing Group and have been also well documented and available in the literature. The works in the literature concluded that security solution in SG requires a holistic method, which combines trusted computing techniques with PKI technologies based on industry standards. In the holistic method, PKI technical elements, such as trust anchor security, attribute certificates, and certificate lifecycle management tools, are the existing technologies tailored specifically to result in an optimal solution for SG networks. To achieve the optimal solution for secure SG networks, the primary step should be taken is to propose a cohesive set of standards and requirements for AMI security.

The authors in [29] articulated the security threats to transmission and distribution (T&D) automation systems. They mentioned that vulnerabilities in power T&D automation systems exist at multiple levels, including components, protocols, and networks. An attack process involves three steps: access, discovery, and control. First, the attacker gains access to the SCADA system through a connection with the corporate network or through a virtual private network (VPN). Subsequently, the attacker studies the behaviors of the system and finally launches an attack. The authors pointed out that the current security solutions are focused mainly on information technology (IT) but not on control systems, and that there are different needs for them, making IT security solutions ineffective. They suggested to decouple the controls from security in order to make it accessible for legacy systems that do not have inherent security. Their work is mainly a conjecture without clear evidence or comparison with other approaches.

5. Challenges and solutions in AMI

Such a complex system undoubtedly presents many challenges. In this section, the challenges and solutions in AMI are identified in two domains including security and communications between networks.

5.1. Challenges and solutions in security of AMI

5.1.1. Challenges

5.1.1.1. Difficulty to identify large-scale catastrophic failures

In AMI security, the primary challenge stems from the high-level dependence between grid components, such that seemingly independent random events can aggregated to yield large-scale catastrophic failures in the grid. High complexity in AMI increases the probability of flaws, and unintended access points increase the possibility of attacks induced failure, especially in an adversary model, in which attacks are readily replicated, thus propagating the failures. In addition, new entities, such as electric vehicles and DER, are expected to be incorporated in the grids. However, researches on security raised up by the incorporations have received very limited attention. Hence, it is very difficult to identify and address the new failure modes in such systems before they become large-scale problems.

5.1.1.2. Dependence between electric grids and communication networks of AMI

We understand the threats to the communication networks of AMI and power grids, and we understand to some extent how the threats associated with the SG communication infrastructure impact on the power grid. However, it is unclear how the threats in the power grids can affect the communication networks of AMI.

5.1.1.3. Challenge to detect network-based threats

The most serious challenge comes from the ubiquitous connectivity in the equipment, software, and controls in AMI. Network-based threats may propagate quickly to overwhelm the whole network of AMI. In addition, the universal connectivity and multiple access points make AMI more vulnerable to attacks (such as DoS). We need to rely on automated detection schemes to respond to network-based threats.

5.1.1.4. Intrusion detection, prevention, and recovery for AMI

Typically, DoS is one of the most dangerous attacks against AMI. If such attack cannot be detected and quarantined early enough, it will risk the failure of the functionality in most critical infrastructure and threaten AMI. Hence, we need new methods for risk assessment based on prior knowledge in order not to introduce further delays in the overall system. Besides, in case that an attack cannot be identified and prevented, appropriate intrusion recovery techniques must be implemented to remedy the consequences of the attack on the critical infrastructure of AMI.

5.1.1.5. Key management techniques for AMI

Today, the majority of key management schemes were proposed only for secure communications within the SG, to address the issues on key establishment for the communicating entities within SCADA systems to protect critical messages, such as near-real-time information, pricing signals, and feedback data regarding energy consumption of customers. In fact, very few studies have been carried out on key management schemes for the AMI. Hence, in the future, researchers should focus on the proposal of novel key management techniques specifically designed for the AMI.

http://dx.doi.org/10.5772/63631

5.1.2. Solutions

5.1.2.1. Security analysis

It is important to develop a risk/security analysis process that can autonomously detect faults to limit the damages to communications of AM. In addition to the analysis of causes and effects of different threats on the electric grid, we need to establish comprehensive failure scenarios that include the impacts of multiple threats simultaneously. The risks include those associated with interactions among cyberspace and physical systems. It will not be possible to consider all possible combinations of threats. Consequently, an automated test system of taking into account different failures (attacks) in both cyberspace and physical systems will be an important additional source for mapping all of the threats and studying their behaviors. Contingency analysis is already performed for analyzing the stability of AMI. However, that will need to be expanded to incorporate the risks due to threats coming from various communication networks in AMI. More precise detection techniques that use multiple factors for accurately predicting threats will need to be devised to reduce false-alarm probability. Based on the previous risk analysis, the algorithms can autonomously detect the faults in AMI to limit the damages caused by degraded security performance.

5.1.2.2. Security standards

On the other hand, international security standards and legislations are also needed for communications in AMI. Currently, there are numerous independent efforts to develop security standards and legislations. Security standards being developed need to be future-proof, considering futuristic applications, operations, and energy markets. Standard test scenarios need to be developed for the researchers developing the algorithms, as well as for equipment manufacturers for detecting security attacks and failure scenarios at the interfaces between power grid and communication networks of AMI. Moreover, we should establish standardized testing requirements for the security in all applications and protocols of AMI. It is also essential to create auditing requirements to ensure compliance with security legislations for utilities, equipment manufacturers, and generators for local, national, and regional regulatory bodies.

5.5.2.3. Quantum key distribution in AMI

The use of quantum key distribution (QKD) can help improve the security of communications in an AMI. Quantum communication is an emerging technology with potential applications to the power grids. QKD has been proposed as an approach to improve the security of communications between the power grids, and it could be implemented over existing fiber-optic channels and free-space optical communication links, within generation systems and power distribution networks. Quantum communication employs a fundamentally different technique from most of traditional communicationtechnologies, and it works based on the physics of entangled quantum states as a fundamental resource. The classical cyber security techniques depend on physical protection of communication channels, and they need complex computational techniques to encrypt transmitted data and protect its confidentiality. The observation of quantum communication measurements fundamentally disturbs the system, alerting the receiver for the changes in the channel. QKD has rapidly matured and is now providing commercial applications by several companies around the world. Researchers are exploring its applications in more challenging and interesting scenarios, including AMI. One potential usage in AMI is quantum location verification. Because today's power system components tend to be stationary, quantum communication techniques could potentially be used to improve the security with regard to the identification of the location of a smart meter. This adds another level of security by ensuring that a smart meter placed at a fixed location in the power grid is truly at that location and is not being spoofed. There are potentially many other applications of quantum communication techniques that might become useful to ensure the security in AMI [1].

5.1.2.4. Cross-layer design for attacks detection

Cross-layer design for attacks detection in communications of AMI based on CR technology is another new research topic. To realize a secure communications of AMI based on CR, security should prevail every other aspects of the whole system design, and be integrated into every system component. AMI security includes the protection of both communication networks and power grids to ensure availability and survivability. The detection techniques based on higher layer introduce an overhead in the network, which could potentially affect timely delivery of critical messages in the SG, resulting in instabilities. Thus, our earlier work proposed a cross-layer design for primary user emulation attacks detection without burdening the networks with extra overhead [30]. In this work, to completely identify primary user emulation attacks and primary users (PU) at PHY layer over multipath Rayleigh fading channels in mobile CR networks, cross-layer intelligent learning capability of secondary user (SU) was exploited to establish radio-frequency fingerprint (i.e., channel-tap power) databases by combining the accuracy and capability of higher layer authentication [31] with a quick detection algorithm on PHY layer [32].

5.2. Challenges and solutions in communications

5.2.1. Challenges

Depending on the characteristics of HAN, NAN, and WAN, different communication technologies are utilized efficiently. For example, in a small area as customers' home, HANs use ZigBee, Bluetooth, or PLC to communicate data between devices. Besides, WiMAX, or WiFi is utilized to build NAN based on wireless mesh topology, and fiber optics or broadband cellular networks are adopted for WANs. However, these traditional communication methods bear the high costs for investment, operation, and maintenance, which are incapable of meeting the requirements and challenges in SG. It has been recognized that CR is a promising technology to construct a more advanced communication infrastructure for SG. By using dynamic spectrum access technique, CR networks solve the problem of scarce spectrum and poor allocation of traditional spectrum policies, and support increasing demand for applications based on wireless communications in SG [33]. In [34], the authors propose the use of CR technology to address the communication requirements, standardization, and security problems of SG communications. There are many benefits brought in by introducing CR into SG. In [35], by using CR technology, it can support energy- and spectrum-efficient designs, as well as avoiding interference and adapting the data throughput, i.e., CR communication over license-free bands is employed in the HANs to coordinate heterogeneous wireless technologies, whereas CR communications over licensed bands is employed in the NANs and WAN to dynamically access unoccupied spectrum opportunities [36].

Moreover, to address aforementioned problems in AMI communications infrastructure (Section 3.2), CR technology can be suitable for AMI communication system. In [37], the authors proposed to enhance a routing protocol for low power and lossy networks (RPL) for CR-enabled AMI networks, i.e., CORPL [38]. This protocol provides novel modifications to RPL to address the routing challenges in CR environments, such as reliable and low latency data delivery, along with protecting the PUs and meeting the requirements of secondary networks. Results show that CORPL improves the reliability of the network while reducing harmful interferences to PUs by up to 50%, as well as reducing the deadline violation probability for delay sensitive traffic. The authors in [39] proposed to use a cloud computing data center as a central communication and optimization infrastructure supporting a CR network of AMI smart meters that is called netbook advance metering infrastructure (Net-AMI). The proposed system is extensible and can easily handle thousands of variations in power systems, communication protocols, control, and energy optimization protocols. By placing new CR antennas on existing cellular antenna towers, vast geographical coverage can be achieved. Moreover, remote software upgrades allow modifications of existing networks components, AMI interfaces, and Net-AMI smart meters in a flexible and amorphous manner using CR technology. In [40], the authors modeled the AMI as a SU in CR-based SG systems based on the IEEE802.22 wireless regional area network (WRAN) [41], which supports the unlicensed operation of SUs with spectrum sensing technologies in VHF/UHF TV broadcast bands from 54 to 862 MHz. The authors also investigated a beam-forming method based on minimum mean squared error (MMSE) to suppress self-interferences in smart meter channels. In [42], the authors proposed a CR-based SG using wireless access communication of line and substation monitoring system addressing the system implementation issues, such as communication efficiency and energy supply in AMI.

As part of the end-user facilities, AMIs can also be efficiently realized with the help of CR technology. For example, by using CR technology, AMI can self-configure and deploy in coexisting wireless networks at various customer premises easily. Based on the spectrum-aware capability of CR, smart meters and equipment in AMI can be easily deployed at the remote sides to achieve reliable and seamless communications between AMIs and the control center of utility company. The cognitive sensor network (CSN) nodes designed with consideration of energy and price limitations in remote monitoring can be the main components for efficient realization of wireless AMI.

However, when we apply CR technology in communications of AMI, we have to face some challenges.

(1) Communications between cognitive HANs and NANs

The challenges to implement communications between CR-based HANs and NANs can be identified as follows.

- **The lack of spectrum holes of licensed bands for data transmissions from smart devices:** In CR-based SG networks, the communications between HANs and NANs are realized by connecting HAN gateways (HGWs) and NAN gateway (NGW). A NGW connects many HGWs from various HANs using licensed bands in an opportunistic manner. However, a SG system generates vast amount of data coming from smart devices. Hence, it might happen that there are no enough spectrum holes of licensed bands to be used for the data transmissions, as there might be times or locations where vacant bands are not available. Moreover, a great challenge in HANs is to internetwork various customer equipment provided by different manufacturers using different standards such as WiFi, ZigBee, WRAN, and Bluetooth.
- **Delay of traffics and real-time capability:** Bidirectional data transmissions between NANs and HANs must meet real-time requirements. The data transmissions involve many types of data, which have different levels of time requirements. For example, the real-time data exchanges between IEDs and other power devices in a large distributed area should ensure that all decisions are made by the control centers in a timely manner, such as controlling or monitoring data, so that demand response can be realized in the customer ends; whereas some other data are transmitted in a periodic manner, such as power consumption data of households. The various types of data also bring in a major challenging issue due to low-speed transmission characteristics and inherent sensing delays of CR. Moreover, the SU in CR must continuously monitor radio spectrum usage to give the precedence to the PU. Therefore, the random interruption of SU traffic will unavoidably cause packet losses and delays in sending SU data. As a result, the communications in a CR network are normally unreliable, and it is a great challenge to support real-time applications.
- **AMI self-configuration:** HANs connect many smart devices to achieve optimum energy consumption and to implement demand response and AMIs. Smart meters, energy management systems (EMSs), and smart devices installed in all customer premises are part of the AMI. The AMI will enable these smart devices to communicate with the utility operated control centers to control their operations at a given time and thus implement demand-side management for the utility. However, the number and characteristics of the smart meters and devices change randomly according to the preferences of customers, who can install new smart meters and devices or remove old smart appliances in an unpredictable manner. Hence, the AMI must have self-configuration ability to ensure online update and effectively monitor the random changes of these smart appliances.

(2) Communications between cognitive NANs and WANs

The challenges to implement communications between cognitive NANs and WANs are identified in the sequel.

- **Limited WAN coverage area due to the use of ISM bands:** The communications between NANs and WANs are built up based on cognitive base stations. Hence, there is also the problem of a shortage of licensed bands for opportunistic access. However, the ISM bands

http://dx.doi.org/10.5772/63631

are not suitable for the communications between NANs and WANs because the coverage area of WAN is larger, whereas the ISM bands are suitable for short-range transmissions.

- **Service reliability using TV white space (TVWS) to connect NANs and WANs:** Another serious challenge using TVWS to connect NANs and WANs is service reliability. In spite of dynamic frequency switching and multi-channel utilization, which can solve the reliability problems, the SU using TVWS is considered as a fundamental issue, in which the SU have to postpone its connections with TVWS if it detects the existence of an incoming PU. New methods to mitigate the unreliability caused by inherent cognitive characteristics of SG communications in the licensed bands remains to be proposed.
- **Scalability:** The scalability feature of WAN connections using wired communication technologies in AMI is limited due to high maintenance and installation costs. Hence, wireless communication technologies are suitable for wide area communications in AMI because of its flexibility. However, to achieve scalability in wireless technologies, we have to add more wireless routers and access points to AMI network, so the installation costs will increase.

5.2.2. Solutions

5.2.2.1. Communications between cognitive HANs and NANs

In order to facilitate the communications between cognitive HANs and NANs, we suggest to use the following techniques.

- **Hybrid spectrum access method to extend the coverage of WANs:** As spectrum holes of the licensed bands may not be enough to transmit a massive amount of data, the communications between HANs and NANs may temporarily operate in the license-free bands (i.e., ISM bands) with lower communication rates. In the method, data transmissions between HGWs and a NGW are considered using a hybrid spectrum access. As a result, the communications between HANs and NANs can improve reliability. As using hybrid spectrum access, the HGWs operate as cognitive nodes in the communication networks and they employ the spectrum sensing method to find vacant spectrum bands. However, if spectrum sensing time of HGWs is too long, then the rest of time to transmit data is short, so the throughput of the networks will be reduced. To solve the problem of HGWs, a scheme to decide when to stop spectrum sensing and when to access ISM bands was proposed in [43] based on the expected throughput performance. In this case, ISM bands are introduced as backup bands for communications to improve service reliability of SG applications. If this condition happens frequently, more NGWs can be installed to utilize space diversity.
- **CR-based AMI self-configuration:** As part of the end-user facilities, AMIs can also be efficiently realized with the help of CR technology. By using CR technology, AMI can self-configure in order to coexist wireless networks at different customer premises. With the spectrum-aware communication capability, smart meters and equipment in AMI can be easily deployed at the remote sides to achieve reliable and seamless communications

between AMIs and the control center of utility company. This is a major opportunity for efficient implementation of wireless AMI in remote monitoring.

5.1.1.1. Communications between cognitive NANs and WANs

To ensure reliable and scalable communications between cognitive NANs and WANs, we identify the approaches as listed in the sequel.

- **WAN coverage area extension to improve reliability:** First, we can use hybrid access modes of licensed and leased bands to extend the coverage area of WANs and improve service reliability. The utilities can lease some radio bands, which are used as backup bands, at a low cost from a telecommunication operator. The hybrid access mode between the leased and licensed bands is intelligently scheduled and seamlessly switched, so that it can improve quality of service (QoS) of data communications, thus benefiting both utilities and users. In this sense, NGWs operate as cognitive nodes, which use spectrum sensing methods to find vacant spectrum bands in communication networks of AMI. After given certain sensing times, the NGWs will choose leased spectrum bands to communicate data with base station, whereas these NGWs still find vacant spectrum bands to use them opportunistically. When the data transmission rate of the leased spectrum bands is higher than that of the cognitive licensed spectrum bands, SUs will stop spectrum sensing and access the leased bands to transmit collected data. On the contrary, if the transmission rate of the leased bands is lower than that of cognitive licensed bands, then SUs will find vacant spectrums and access the cognitive licensed bands to transmit data for achieving a higher throughput. However, the number of the leased spectrum bands is very limited and they also serve as backup bands in emergency situations to transmit critical data. Hence, the NGWs have to spectrum sensing periodically to release the leased spectrum bands once a vacant spectrum band is identified. In NANs, the available unoccupied spectrum bands are scarce in urban areas, whereas abundant in rural areas, because the amount of data traffic in urban areas is much larger than that in rural areas. Therefore, the leased bands, which are distributed to a NAN in urban areas, should be more than that distributed to a NAN in rural areas. Moreover, a leased spectrum band can be shared by several NANs without causing interference to each other if the service area of a WAN is very large. Similarly, the leased bands are utilized as backup bands for communications to improve service reliability of SG applications. Besides, we can use cooperative communications to extend the coverage and improve service reliability. Other available wireless and wireline technologies, such as wireless cellular networks, internet, and fiber optics, should also interoperate with cognitive NANs and WANs to make SG more resilient, scalable, and reliable in an economical manner. For example, currently, the mobile communications have been implemented via both cellular networks and IPv6 mobile ad hoc networks (MANETs), so that we can utilize the MANETs to transmit noncritical data.

- **Scalability:** The CR technology creates an opportunity to increase scalability with a low cost. For example, IEEE 802.22 standard has unique features, such as geo-location, spectrum sensing, and intra-system coexistence for CR-based operations. The standard operating in TVWSs from 54 to 862 MHz permits broadband wireless access to wide range rural areas without interference with the PUs. Using IEEE 802.22 standard, coverage area of base station can be 33 km if customer premises equipment operates at power level of 4 W. When higher power levels are allowed, the coverage area can be increased to 100 km [34].

6. Conclusions and future visions of AMI

AMI based on smart meters in SG has been identified, and their state-of-the-art research activities were reviewed. In addition, the issues on security of AMI in SG have also been discussed. Future SG should comprise intelligent monitoring systems to keep a track of all electricity flows and a huge amount of collected data from smart devices as well. Hence, it must be flexible and resilient to accommodate new requirements in an economical manner. To achieve these goals, communications in AMI based on CR will certainly play an important role for infrastructures of the SG. Moreover, by using AMI, SG can support real-time traffic delivery with stringent the quality of service requirements of real-time applications. The major challenges on evolutional path toward SG and solutions are also identified in this chapter. With AMI, SG should preserve its interoperable and secured communications within a hybrid system where both new and legacy grids coexist. Therefore, AMI in SG should be built up on open protocols with a common notion of security and standard. Besides, advanced research topics, such as artificial neuron network and Fuzzy theory, can also apply to the intelligent monitoring systems to improve the ability of AMI. Moreover, accurate state estimation methods need to propose in the future to detect blind false data injection attacks because accurate state estimation is of paramount importance to maintain normal operations of AMI. Typically, a bad data detection system is used to ensure the integrity of state estimation and to filter faulty measurements introduced by device malfunctions or malicious attacks. However, in [44], we prove that blind false data injection attacks using the principal component analysis approximation method without the knowledge of Jacobian matrix and the assumption regarding the distribution of state variables can bypass the bad data detection system to inject fault data in the system. In the future, the architecture of AMI not only aims at seamless integration of various existing smart metering products, but also other software systems used by power utilities (i.e., outage, energy and distribution management systems, etc.). Hence, new solutions aim at enabling flexible integration of metering devices and their grouping in form of *virtual meters* through hierarchically organized software structures and flexible standardized communication interfaces.

Author details

Trong Nghia Le[1], Wen-Long Chin[2*], Dang Khoa Truong[1] and Tran Hiep Nguyen[1]

*Address all correspondence to: wlchin@mail.ncku.edu.tw

1 Le Quy Don Technical University, Hanoi, Vietnam

2 National Cheng Kung University, Tainan, Taiwan, ROC

References

[1] IEEE Smart Grid Research. IEEE Vision for Smart Grid Communications: 2030 and Beyond. IEEE Vision for Smart Grid Communications:. May 2013; Chapter 1; 3–22.

[2] V.C. Gungor, B. Lu, and G.P. Hancke. Opportunities and challenges of wireless sensor networks in smart grid. IEEE Trans. Ind. Electron. Oct. 2010;57(10):3557–3564.

[3] U.S. Department of Energy. [Online]. Available at: http://www.oe.energy.gov. 2011.

[4] A.Y. Saber and G.K. Venayagamoorthy. Plug-in vehicles and renewable energy sources for cost and emission reductions. IEEE Trans. Ind. Electron. Apr. 2011;58(4):1229–1238.

[5] P. Siano, C. Cecati, C. Citro, and P. Siano. Smart operation of wind turbines and diesel generators according to economic criteria. IEEE Trans. Ind. Electron. Oct. 2011;58(10): 4514–4525.

[6] D.G Hart. Using AMI to realize the smart grid. in Proc. IEEE Power and Energy Society General Meeting - Conversion and Delivery of Electrical Energy in the 21st Century. July 2008;:1–2.

[7] R. Gerwen, S. Jaarsma, and R. Wilhite. Smart metering. [Online]. Available at: http://www.leonardo-energy.org/webfm_ send/435.

[8] S.S.S.R. Depuru, L. Wang, and V. Devabhaktuni. A conceptual design using harmonics to reduce pilfering of electricity. IEEE PES General Meeting; Minneapolis; MN. July 2010; 1–7.

[9] R.R. Mohassel, A. Fung, F. Mohammadi, and K. Raahemifar . Application of Advanced Metering Infrastructure in Smart Grids. 2014 22nd Mediterranean Conference on Control and Automation (MED); Palermo. 2014;:822–828.

[10] Silicon Laboratories, Inc. Smart Metering Brings Intelligence and Connectivity to Utilities, Green Energy and Natural Resource Management. Rev.1.0. Available at: http://www.silabs.com/Support%20Documents/TechnicalDocs/Designing-Low-Power-Metering-Applications.pdf. August, 2013.

[11] R. Berthier, W. H. Sanders, and H. Khurana. Intrusion detection for advanced metering infrastructures: Requirements and architectural directions. *2010 First IEEE International Conference on Smart Grid Communications (SmartGridComm); MD*. 2010;:350–355.

[12] R. Shein. Security measures for advanced metering infrastructure components. in Proc. *2010 Asia-Pacific Power and Energy Engineering Conference* (APPEEC); Chengdu. 2010;:1–3.

[13] S. Rusitschka, K. Eger, and C. Gerdes. Smart Grid Data Cloud: A Model for Utilizing Cloud Computing in the Smart Grid Domain. First IEEE International Conference on Smart Grid Communications; Gaithersburg, MD. 2010;:483–488.

[14] Anderson, C. Zhao, C. Hauser, V. Venkatasubramanian, D. Bakken and A. Bose. A virtual smart grid. IEEE Power & Energy Magazine. December 13, 2011;9(6):49–57.

[15] D. Backer. Power Quality and Asset Management The Other "Two-Thirds" of AMI Value. 2007 IEEE Rural Electric Power Conference; Rapid City, SD. May 2007;:6–8.

[16] S. W. Luan, J. H. Teng, S. Y. Chan, and L. C. Hwang. Development of a smart power meter for AMI based on zigbee communication. *2009 International Conference on Power Electronics and Drive Systems (PEDS); Taipei*. 2009;:661–665.

[17] W. Luan, D. Sharp, and S. Lancashire. Smart grid communication network capacity planning for power utilities. *IEEE PES T&D 2010*; New Orleans, LA, USA. 2010;:1–4.

[18] B. Reid. Oncor electric delivery smart grid initiative. *2009 62nd Annual Conference for Protective Relay Engineers*; Austin, TX. 2009;:8–15.

[19] R.R. Mohassel, A. Fung, F. Mohammadi, and K. Raahemifar. A survey on advanced metering infrastructure. Int. J. Electr. Power Energy Syst. Dec. 2014;63:473–484.

[20] Z. Wan, G. Wang, Y. Yang, and S. Shi. SKM: scalable key management for advanced metering infrastructure in smart grids. IEEE Trans. Ind. Electron. Dec. 2014;61(12): 7055–7066.

[21] Murrill BJ, Liu EC, Thompson II, RM.. Smart Meter Data: Privacy and Cyber security. Congressional Research Service; USA. 2012; 1–45.

[22] Molina-Markham A, Shenoy P, Fu K, Cecchet E, Irwin D. Private memoirs of a smart meter. Proceedings of the 2nd ACM workshop on embedded sensing systems for energy-efficiency in building; ACM, 2010; 61–66.

[23] G. Kalogridis, C. Efthymiou, S.Z. Denic, T.A. Lewis, and R. Cepeda. Privacy for smart meters: towards undetectable appliance load signatures. In: Proc IEEE International Conference on Smart Grid Communications, Gaithersburg, MD. 2010.

[24] Pfitzmann A, Hansen M.. A terminology for talking about privacy by data minimization: anonymity, unlinkability, undetectability, unobservability, pseudonymity, and identity management. 2010. Online: http://dud.inf.tu-dresden.de/Anon_Terminology.shtml

[25] The Advanced Security Acceleration Project. Security profile for advanced metering infrastructure. AMI-Sec Task Force, TN, Tech. Rep. Jun. 2010.

[26] N. Liu, J. Chen, L. Zhu, J. Zhang, and Y. He. A key management scheme for secure communications of advanced metering infrastructure in smart grid. IEEE Trans. Ind. Electron. 2013;60(10):4746–4756.

[27] M. Nabeel, S. Kerr, X. Ding, and E. Bertino. Authentication and key management for advanced metering infrastructures utilizing physically unclonable functions. Purdue University, Tech. Rep. 2012.

[28] A.R. Metke, and R.L. Ekl. Security technology for smart grid networks. IEEE Trans. Smart Grid. 2010;1(1):99–107.

[29] D. Wei, Y. Lu, P. Skare, M. Jafari, K. Rohde, and M. Muller. Power infrastructure security: fundamental insights of potential cyber attacks and their impacts on the power grid. Part of project Protecting Intelligent Distributed Power Grids against Cyber Attacks for DoE: 35, 2010. [Online]. Available at: http://citeseerx.ist.psu.edu/viewdoc/summary?doi=10.1.1.152.1832.

[30] T.N. Le, W.L. Chin, and W.C. Kao. Cross-layer design for primary user emulation attacks detection in mobile cognitive radio networks. IEEE Commun. Lett. May 2015;19(5):782–799.

[31] C. N. Mathur, and K. P. Subbalakshmi. Digital signatures for centralized DSA networks. *2007 4th IEEE Consumer Communications and Networking Conference*, Las Vegas, NV, USA. 2007;:1037–1041.

[32] W.L. Chin, T.N. Le, C.L. Tseng, W.C. Kao, C.S. Tsai, and C.K. Kao. Cooperative detection of primary user emulation attacks based on channel-tap power in mobile cognitive radio networks. Int. J. Ad Hoc Ubiquitous Comput. 2014;15(4):263–274.

[33] J. Wang, M. Ghosh, and K. Challapali. Emerging cognitive radio applications: a survey. IEEE Commun. Mag. Mar. 2011;49(3):74–81.

[34] A. Ghassemi, S. Bavarian, L. Lampe. Cognitive Radio for Smart Grid Communications. First IEEE International Conference on Smart Grid Communications (SmartGridComm); Gaithersburg, MD. 2010;:297–302.

[35] Vineeta and J. K. Thathagar. Cognitive radio communication architecture in smart grid reconfigurability. *2012 1st International Conference on Emerging Technology Trends in Electronics, Communication and Networking (ET2ECN)*; Surat, Gujarat, India. 2012;:1–6.

[36] Y. Han, J. Wang, Q. Zhao, and P. Han. Cognitive information communication network for smart grid. *2012 IEEE International Conference on Information Science and Technology*; Hubei. March 2012;:847–850.

[37] A. Aijaz, H. Su, A.H. Aghvami. CORPL: a routing protocol for cognitive radio enabled AMI networks. IEEE Trans. Smart Grid. Jan. 2015;6(1):477–485.

[38] T. Winter. IPv6 routing protocol for low power and lossy networks. Internet Engineering Task Force, RFC 6550. Mar. 2012. [Online]. Available at: http://www.rfc-editor.org/rfc/rfc6550.txt.

[39] K. Nagothu, B. Kelley, M. Jamshidi, A. Rajaee. Persistent NetAMI for microgrid infrastructure using cognitive radio on cloud data center. IEEE Syst. J. 2012;6(1):4–15.

[40] S. Chang, K. Nagothu, B. Kelley, M.M. Jamshidi. A beamforming approach to smart grid systems based on cloud cognitive radio. IEEE Syst. J. 2014;8(2):461–470.

[41] IEEE Draft Standard for Information Technology - Telecommunications and information exchange between systems - Wireless Regional Area Networks (WRAN) - Specific requirements - Part 22: Cognitive Wireless RAN Medium Access Control (MAC) and Physical Layer (PHY) Specifications: Policies and Procedures for Operation in the TV Bands. IEEE P802.22/D3.0. March. 2011.

[42] B. Li, B. Zhang, J. Guo and J. Yao. Study on Cognitive Radio Based Wireless Access Communication of Power Line and Substation Monitoring System of Smart Grid. 2012 International Conference on Computer Science and Service System (CSSS); Nanjing. 2012;:1146–1149.

[43] F. Liu, J. Wang, Y. Han, P. Han. Cognitive radio networks for smart grid communications. 9th Asian Control Conference (ASCC); Istanbul. 2013;:1–5.

[44] Z.H. Yu and W. L. Chin. Blind false data injection attack using PCA approximation method in smart grid. IEEE Trans. Smart Grid. May 2015;6(3):1219–1226.

Chapter 4

The Post Carbon City and Smart Metering

Andrew Haning and Martin Anda

Additional information is available at the end of the chapter

http://dx.doi.org/10.5772/64082

Abstract

Buildings and districts are an appropriate focus for smart metering infrastructure in the urban environment. While properties and buildings have traditionally been metered for revenue recovery purposes, energy management of these buildings has not been available. In the way we account for money, we should account for energy; energy in its own right carries a direct cost with it to the end user. Along with carrying a cost, energy also carries carbon emissions. Smart meters are a vital component to the making and management of post carbon cities and can be used to monitor not only electricity use but also water and gas consumption. Energy Management Systems combined with structured metering also enable consumers with renewable energy generation such as photovoltaic (PV) panels to monitor their own generation, consumption, import and export. As battery storage becomes integrated with renewable energy generation, consumers will have the ability to consume cheaper renewable energy than can be bought from the grid and sell energy back to the grid at the most economically viable times. While uncertainty surrounds the grid and its impact on rising electricity prices, smart metering, intelligent control systems and utilities offering consumers more amenities and the ability for consumers to participate in the wholesale market will ensure the smart grid can contribute to future carbon neutral urban environments.

Keywords: smart meter, smart grid, post carbon city, renewable energy, energy use, water use, buildings, districts

1. Smart meters

Traditionally meters have been used for revenue recovery on utility scale distribution networks with meters read manually on a monthly basis. Meters and data only accounted for a single direction of energy flow. Smart meters measure a spectrum of energy consumption information at intervals and communicate the information to the utility remotely. They are effective tools for gathering of data, monitoring long-term trends and responding short term or live events or incentives mechanisms such as demand-side management both locally and remotely. New technology in smart meters enables the measurement of two-way energy flows [1]. Monitoring of data allows for a higher degree of control on energy use, which will be important in the emerging post carbon urban environments [2] with the impact of distributed renewable energy generation, battery storage and electric vehicles as well as feedback on energy use patterns to consumers.

Buildings and districts are an appropriate focus for smart metering infrastructure in the urban environment. While properties and buildings have traditionally been metered for revenue recovery purposes, energy management of these buildings has not been available. In the way we account for money, we should account for energy; energy in its own right carries a direct cost with it to the end user. Along with carrying a cost, energy also carries carbon emissions. Smart meters are a vital component to the making and management of post carbon urban environments [3].

Smart meter requires accessible communications for robust data availability and storage to enable advanced architecture and applications. Preferably, it is best to use a smart meter with multiple communication ports (at a minimum Modbus, TCP/IP, IR). Smart meters need to be installed and calibrated to national peak measurement body standards (e.g. in Australia, the National Measurement Institution or NMI) to enable utility billing. The smart meter should have a configurable program to optimise data measured and collected to meet the requirements of application (i.e. within the South West Interconnected System (SWIS), demand intervals have been extended from a 15-min period to a 30-min period).

The meter should have the ability to bring in additional data sources (i.e. water, air, gas, electric and steam, WAGES) through either direct pulse inputs or through a local wireless communication protocol such as Zigbee and its appropriate sub-meter hardware. In addition, smart meters should have configurable on board logging that enables backup of meter data locally if communications or power failures occur, which can cause loss of data and revenue. Meters and data collection needs to adhere to legislation on data collection and usage. Consumers' data are typically protected, and permission is required to use it (check your local laws and regulations to ensure compliance).

In this chapter, we will describe smart meter architecture and management systems for buildings and district energy and water use, and then, we will provide two case studies as to how this is applied in practice. The latter is done with specific industry focus and products.

2. Smart meter architecture

Smart meter architecture needs multiple levels providing multiple points of reconciliation. Hierarchies should inform the levels of consumption and more importantly demand within the distribution network and for total and isolated services (i.e. Internal/External Lighting, Mechanical services, power etc.). Smart meters should separately meter all generation inputs into the system (i.e. renewable energy, Co/Tri-generation). Using hybrid communication architecture will bring in meters at the lowest cost and highest data reliability. (Direct meter to TCP/IP; Meter via Modbus to gateway to TCP/IP; Gateway to enable localised data logging [where proprietary meter software is not used to pull data logs direct from the meter] and pass through of centralised/decentralised [i.e. cloud based] software packages to contact meters/ gateway to collate data).

Accuracy of data is important for smart metering metrology characteristics. International standards assist to ensure accurate measurements. Varying reference conditions such as low loads, poor power factor and/or verification of meter accuracy affect performance outcomes. Accuracy standards such as Class 0.5S to AS/IEC 62053-22 improve levels of reliability thus ensuring confidence in measurements in varying conditions.

The use of smart meters can leverage off existing WiFi technology networks to reduce costs. GSM technology is expensive and has ongoing running costs (bring data back over TCP/IP backbone; and architecture complies with BEEC/NABERS & Green Star/BREEAM/LEED) (**Figure 1**).

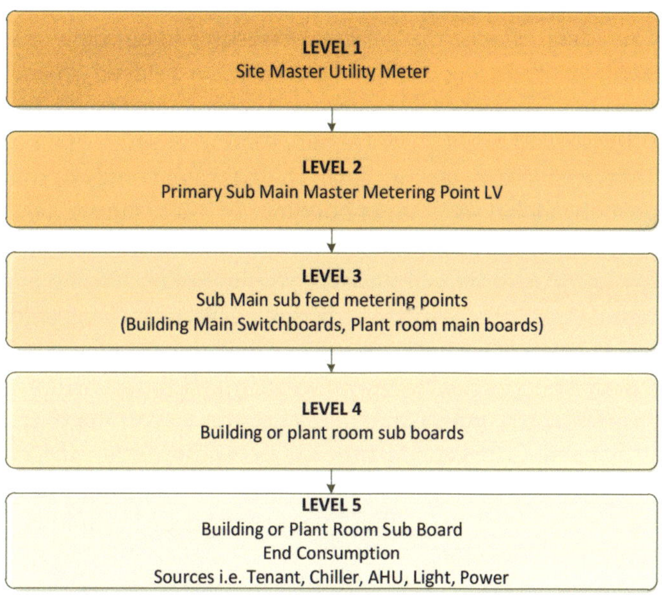

Figure 1. Large building and campus electricity smart meter architecture.

3. Energy management system (EMS)

Centralised energy management systems (EMSs) are used to support the reporting of energy consumption monitoring, a primary requirement as a component of the management cycle for feedback and continual improvement. EMSs directly support the billing and revenue recovery from end-use consumers. It allows customers to understand where the costs are allocated from their bill and therefore where energy and cost savings can be made. EMSs assist the implementation of demand-side management, often with alarms linked to either a manual intervention or more preferable automated actions implemented through the building management system (BMS).

Demand and particularly the capacity charge of the bill is the most effective way to reduce energy consumption (predictive based upon historical consumption and ambient conditions particularly for the Perth climate as HVAC often accounts for ~40% of a sites consumption). EMSs can set benchmarks for energy savings, enable forecasting, ensure benchmarks are achieved and energy efficiency measures are tracked and continue to perform as installed and commissioned.

Dashboards are typically used as a primary method to advise building occupants on a live basis when demand-side management is in effect and what measures are being undertaken. This is particularly important to manage stakeholder expectations during these critical demand management events where large amounts of money can be saved on capacity charges. The capacity charges are price signals from the utilities to reduce peak demand of the site when the grid is at maximum demand.

There are several key demand-side management measures to be communicated to building occupants. One example is reducing the lighting demand in a staged process, with the lowest level not to significantly affect work and in line with occupational health and safety. Another example is expanding set point ranges for thermal comfort to reduce pressure on the chilled water supply and the need to turn on additional chilled water supply to cope with escalating demand. A change in the global set point of buildings of 2–3°C during demand events when the external ambient temperature is >38°C for a period of 2–5 h would likely not negatively affect occupant comfort. An additional measure would be the shutting of a building CHW supply or air-conditioning entirely for 30 min periods. The Australian National University in Canberra using SATEC meters have successfully trialled this technique to manage the load. They advised that ambient temperatures did not escalate and stayed within 1°C. An important benefit is reduced fan and pump power and energy use of mechanical services. Buildings could run on a cyclical demand management process, depending upon the event in question. Another example is the pre-cooling of buildings before 8 am during periods of consecutive 38°C ambient temperatures. (See the CIBSE Energy Efficiency in Buildings for more tested ideas as well as for pros and cons) [4].

4. Integrated and distributed renewable energy systems in buildings

Smart meters are integral to integrated and distributed renewable energy systems in buildings. If an existing building is considering renewable energy, smart meter data can be analysed to size the system to produce the greatest benefits to the owner/occupier. Demand profile in the form of 15 or 30 min interval data can be used to determine seasonal consumption patterns, base and peak loads. Renewable energy production can then be modelled on an equivalent 15 or 30 min interval data and compared on a temperature and humidity normalised basis to the predictive model to gauge the optimal pairing of renewable production to consumption behaviour.

While smart meters provide accurate data, in general unless the occupier is planning on going carbon neutral or implementing battery storage, renewable energy should only ever be sized to 70–75% of average daily load on the normalised comparison data. This provides a buffer to enable future energy efficiencies to reduce the overall consumption of the premises. However, this needs to be optimised with the degradation of PV and other renewable energies over the life cycle of the system, typically 20–30 years for PV.

With the introduction of energy storage, buildings are now able to more effectively utilise their renewable energy production and target energy generated to best meet their own needs. The largest benefit to large consumers is the ability to target market signals such as demand-side management programs and to reduce their own capacity charges. Other benefits offered include the ability to store the energy rather than export it to the grid often for little or no rebate from the utility supplier. Meters form an integral part in the monitoring and management integrated renewable energy and storage systems. Live data inform all decision making and supports tuning of these management systems to improve performance and reduce both carbon emissions [5] and consumers' bills.

5. Residential smart metering

Traditional residential metering has involved an electricity, gas and water meter, which is read on a monthly, bimonthly, quarterly or biannual basis. This standard of data does not provide sufficient detail of consumption patterns to enable energy or water management [6]. It is vital that utilities and embedded network owners provide consumers access to live and/or 15–30 min interval data [7]. The data should be easily accessible either by an in-home display (IHD), mobile application or website [8].

Consumers are then informed on a real-time basis of how their behaviour contributes to their utility bills and are empowered to manage their own utility bills [9]. In Australia, residential electricity prices have increased 60–80% over the past 10 years, which has put extra additional strain on consumers' budgets, in particular pensioners and low-income households, the most vulnerable to fluctuations in the community. Data should be easily accessible and easily configurable at the tip of a finger for consumers, particularly for those who are not technology literate or who do not have access to the internet, a computer or smart phone. In this latter case, affordable IHDs would be the best solution [8].

Utilities and energy service companies can then configure the data to provide useful feedback to consumers in water and energy efficiency and savings programs. This can be done through the IHDs, online programs, smart phone apps or over-the-phone coaching programs [3].

The typical structure, applications and interactions of residential smart metering are laid out in **Figure 2**.

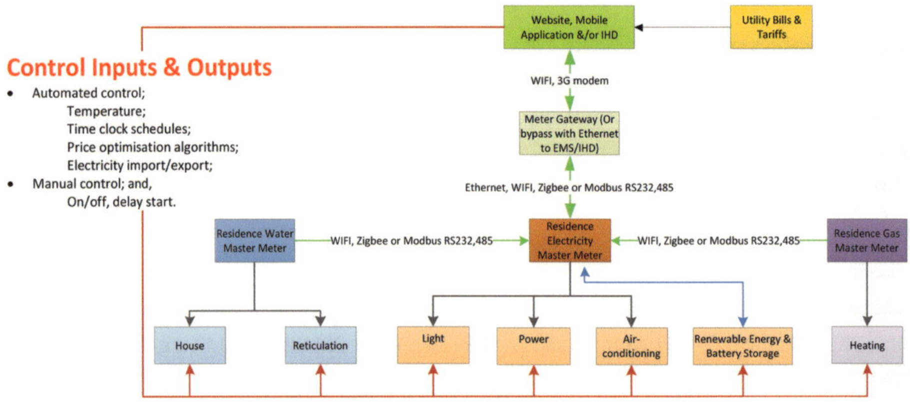

Figure 2. Typical residential smart meter architecture and applications.

6. Case studies

In this section, we introduce two case studies that show how the above principles are applied in practice at two very different locations. Case study 1 is in Perth, Western Australia on the Murdoch University Campus where the Carbon View Smart Meter Project was implemented and provides a commercial/institutional example. The second case study is described in Central Park, Sydney, New South Wales, Australia. This is an inner city mixed use development and this provides a residential apartment example.

In both case studies, the same types of meters have been used, being the SATEC EM 133 smart meter, in very different applications. The difference at each site is the programming. Although at Murdoch University the SATEC meters were programmed for a 15-min demand interval and one decimal place (to account for large users), and the meters at Central Park were programmed for a 30-min demand interval and four decimal places for the smaller residential users. For example, the latter provides significant figures for the water meters down to liters. This is specifically because of the type of pulse output from the water meters at this site.

6.1. Case study 1: Murdoch University campus carbon view smart meter project

Murdoch University has approximately 135,000 m^2 gross floor area (GFA) across 165 buildings on the South Street Campus. The university distributes electricity from the western power grid through two gate feeders through two interconnected HV ring mains at 22 kV with no step down. Each of these independent feeders is metered by western power. The internal distribution network is broken down into two distinct electrical ring mains with a total of 26 substations, which reduce HV (high voltage) to LV (low voltage). The locations and layout of the HV distribution network can be seen in **Figure 3**. Electricity is then reticulated out to the end use either directly or through building's internal LV distribution network at 240/415 V.

http://dx.doi.org/10.5772/64082

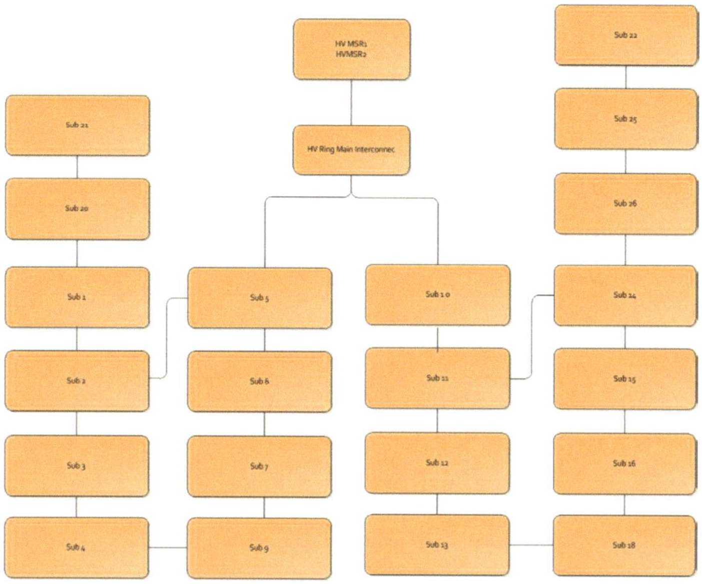

Figure 3. Murdoch University Typical HV Schematic Distribution Single Line Diagram, each ring main has an open point not represented in the diagram which is moved to balance the ring and to enable maintenance.

Murdoch University has recently embarked upon a smart meter roll out to upgrade its existing varied and aging meter embedded network infrastructure. Murdoch University undertook a review of its electricity metering embedded network and found that its old mechanical meters were at end of life and required replacement. After external consultation by JD Shute Pty Ltd, the SATEC EM133 meter was chosen as the universities' meter of choice for all installations as a standardised product. The university initially purchased 200 m to be rolled out from 2015 to replace existing end of life metering which included central chilled water plant and major tenancies.

Figure 4. The SATEC EM133 meter.

A significant amount of work was put in through the consolation phase on the meter selection to ensure that the meter had the capability to meet the needs of the university. One of the primary governing factors as to the choice of the SATEC smart meter was the need for a NMI-approved DIN-rail mount meter. At the time of selection, the SATEC EM133 meter was the first meter on the market available, which met the criteria [10].

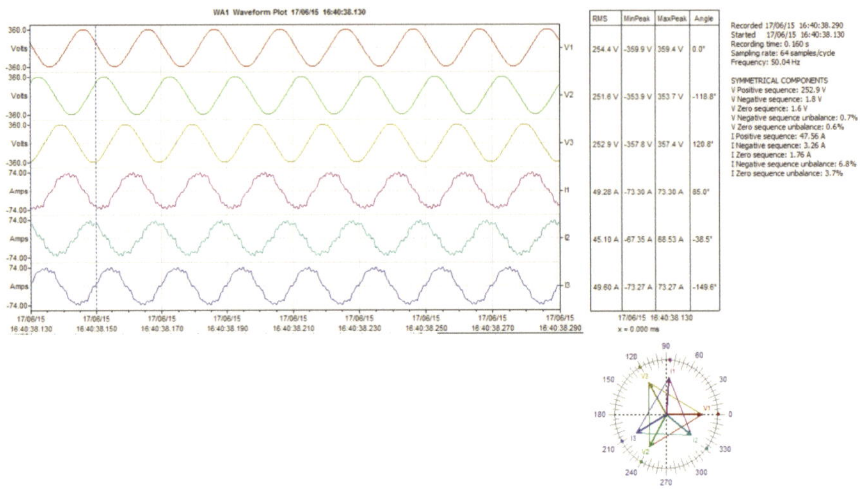

Figure 5. Instantaneous sub-cycle meter analysis and phase rotation vector diagram.

The SATEC EM133 meter, **Figure 4**, provides the university with flexibility as to what information can be programmed, stored and logged locally and at what frequency as well as live Modbus values, which can be pulled from the meter remotely by an EMS. The SATEC meter is provided with free software that provides access to program, interrogate and download data logs. This licence-free power analysis software (PAS) enables a sub-cycle snapshot and phase diagram to be produced (**Figure 5**), which is very useful for commissioning and electrical investigations.

Figure 6. The variety of existing meters installed at Murdoch University.

http://dx.doi.org/10.5772/64082

It is recommended that within DBs that 0–100Amp direct meters are placed between the main switch and the chassis or end use rather than on the source side of the main switch as the entire DB including the other services will have to be isolated should the meter fail or require replacement. Therefore, this configuration of direct connect in line metering significantly reduces the impact and stakeholder engagement for facilities managers during shutdowns.

The alternative CT type meters while they take up a significantly larger amount of space within the DBs due to the installation with Class 0.5 s CTs allow for installed test links to be isolated and the power from the source to the measured load to remain on while the meter is being replaced. CTs are also required for measurement of loads >100 A. The university's choices of CTs for CT type installations are the:

- Type 'S', 200/5 Amp CT 5VA Burden SCT200-IPD; and,
- Type 'T', 800/5 Amp CT 15VA Burden TCT800-IPD.

These CTs are Class 0.5ME2 and 0.5S extended range CTs, which enable the SCT200 CTs to be installed on measurement application up to 400 A loads and the TCT800 CTs to be used on applications up to 1600 A. With the use of extended range CT's, the smart meters are required to be of the extended range design. SATEC's EM133 supports extended range CT's thus compliance to 5A/10A style CTs is assured. The majority of metering hardware is designed as 5A/6A which would not be suitable for extended range CT's.

The third alternative installation methodology used at the university is the SATEC EM 133 meter installed with high-accuracy current sensors (HACS) as the installed equipment is NMI compliant as a Class 1 system. This meter configuration has only recently become available since NMI installation and provides the additional benefit to the university particularly with the ability to use HACS split core CTs on existing electrical infrastructure rather than having to disconnect to slide the IPD CTs over the cables and then re-terminate the cables. The HACS CTs range from 5 to 1200 A with various split core and solid core configurations. HACS CT installations with the EM 133 meter are NMI compliant up to 200 m with the CT wiring only transmitting 40 mA.

When developing a metering architecture and strategy for the existing campus building, it was important to be flexible yet also future proof the strategy for the expansion within the campus. While a significant amount of planning and stakeholder engagement with academics and industry experts took place, changes to the base program still occurred post-installation. The original program had four decimal places and was meant for residential applications, as the base program for the SATEC EM133AR was designed for the Central Park installation measuring water down to 0.0001 kWh intervals. This problem was noticed when regular meter readings of large consumers were lower than the previous read. It was then discovered through the use of the 30-min interval logging that the meter had been clocking at 9999.9999 kWh, which in some instances was every few days. Thanks to historical log files because no data were compromised nor lost in the process.

The MU program has since been tailored for commercial applications and reduced to 1 decimal place. As a result, the largest consumer on campus will clock in around 25 years time rather than have constant clocking in some cases every 2–3 days with large consumers. Choosing this number of decimal places does impact pulse counted units such as water and gas. Water will, therefore, have the lowest resolution down to 0.1, which will equate to 100 L. Gas likewise will clock at 0.1 of an m^3.

Investigations were undertaken to determine the most cost effective and robust connection for the automated meter reading software. Initially, the plan to bring metering online to the BMS utilising the BMS backbone through upgraded PLC in the form of Schneider Automation Servers was not feasible as there was a limitation to the number of Modbus points that could be logged. Another finding was that the automation server would not function as a gateway device to allow other software packages to contact the meter. Therefore, the license free and very useful commissioning and meter management tool PAS could not be used remotely for commissioning, meter data log reclamation, programming and detailed engineering analysis through its sub-cycle waveform capture.

The university meter program supports the functions in **Table 1** in relation to the eight assignable registers within the meter as well as providing time-of-use (TOU) data in line with the Western Australian SWIS grid on peak and off peak tariff times. That TOU billing is represented below:

- On peak—Monday to Friday 8 AM to 10 PM—Tariff 2 (displayed as T2 on meter display); and
- Off peak—Weekends and after hours—Tariff 1 (displayed as T1 on meter display).

The information in **Table 1** is also represented on the meter display and logged daily at the following configurations. Current totals for the target registers are displayed in the meter for all time. The previous and second previous read values at the end of the daily, weekly, monthly and quarterly period are displayed along with the period use consumption values for each target register as sub-menus on the meter display.

Meter program registers	Unit	Source input	Target
Total electricity	kWh	kWh import	Register 1
Phase 1	kWh	kWh L1 import	Register 2
Phase 2	kWh	kWh L2 import	Register 3
Phase 3	kWh	kWh L3 import	Register 4
Kvarh	Kvarh	Kvarh import	Register 5
kVAh	kVAh	kVAh import	Register 6
Gas	m^3	DI1	Register 7
Water	m^3	DI2	Register 8

Table 1. The Murdoch University PAS meter program for assignable registers and time of use logging (TOU).

TCP/IP was then considered against a typical Modbus RS485 meter communication connection to a gateway. While the university would have liked to go direct with the SATEC meter TCP/IP module onto its fiber backbone, the cost per module and per switch was high enough for the university to choose to utilise a hybrid solution to maintain cost-effectiveness across the smart meter roll out.

The university chose the Schneider Com X 510 gateway, which can log up to 30 m, but has been limited to around 12 m due to the number of Modbus points logged and the high polling frequency required. The university is still working on finding the limit to the number of meters per gateway and total distance for RS485 to optimise cost-effectiveness and ensure that live polling is equal to or less than 500 ms. This is particularly the case as the SATEC EM 133 driver being viewed and logged on the Schneider Power Manager Software has not been undertaken before. One of the biggest benefits of choosing the Schneider Com X 510 gateway other than the competitive price was that the software is intrinsically linked and viewed on MU's smart structure building operations (SBO) despite being installed on a standalone server and on a secure VLAN. Therefore, SBO works as a high-level Web integration package for the BMS, EMS and, in the future, the lighting controls of the university, which will all be integrated into a demand-management page on the BMS.

Figure 7. Installed Embedded Network Installations in 2015 at Murdoch University: from top left and left to right each row (a) 9× SATEC EM133 meters installed on main distribution board in one building, (b) chart showing output from the Schneider Electric COM510 energy server webpage, (c) showing the 3× energy servers with ethernet and modbus connections, (d) and (e) showing the EM133 smart meter installation using IPD CTs, test links and voltage fuses.

Building design and specification of meters and electrical installations have varied over 40 years as building codes have changed leading to variation in metering installed across the campus, which can be seen in **Figure 6**. Within buildings, electrical risers, disparate DBs and in slab conduit can limit the ability to run Modbus RS485 cable in an open loop to the nearest Com X 510 device. As a result, meters in this situation are installed with the SATEC EM133 Ethernet module, which can leverage off of the universities' Ethernet network; the Ethernet module that 'bolts' onto the side of the SATEC EM133R meter. While this module increases the installation cost of the meter significantly, the meter can still be logged at a high level either by the Com X 510 or at the EMS itself using 4 TCP Sockets. In some locations, the Com X 510 gateways have leveraged off of the universities' *Eduroam* WiFi network to bring remote sub-stations, which are not connected by the universities' fibre backbone, such as sub-station 18 in **Figure 3** to provide connectivity and visibility on the EMS. Utilising the universities' WiFi network for remote sites has resulted in significantly reduced costs to bring these integral parts of the embedded network online. If there will be loss of communications, the biggest benefit of the Com X 510 is that it will log approximately 45 Modbus points at 1 min intervals; these data can then be recalled by the EMS automatically when connectivity is resumed. Installed Com X 510 devices connected through Ethernet are visible in **Figure 7**.

PV and renewable energy metering applications will always be connected through the Ethernet module to ensure the highest resolution for research and to ensure automated protection mechanisms are not affected by network traffic or WIFI outages.

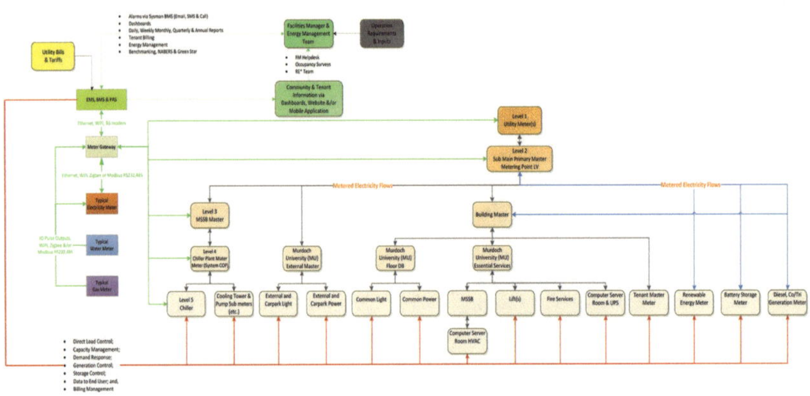

Figure 8. Typical Murdoch University Embedded Meter Network SLD, automated meter reading system and theoretical energy management framework. The arrows depict metered energy flows and type. The 5 level and end use breakdown is important for energy management, NABERS and Green Star.

The embedded network meter hierarchy in **Figure 1** has been adapted and represented along with the individual SLD-type diagram in **Figure 8**, which also displays the communications methodology in place at Murdoch University. The embedded network diagram also shows the breakdown of end use services in line with NABERS and Green Star requirements as well as the interactions between the meters, the EMS and the BMS to form the management feedback loop for the control of consumption and internal and external conditions for occupation and operation of the facility.

The outcome is the system that has live consumption information, internal building feedback from equipment and sensors and building and tenant occupant feedback and room booking information. The deployment of sensors as part of this system must be done in a thoughtful and strategic manner to satisfy feedback requirements and to optimise data collection for this purpose [10]. In the end, this has leveraged existing systems to enable the energy and water efficient operation of the Murdoch University facilities at the lowest operational cost and environmental impact.

6.2. Case study 2: Central Park, Sydney, an inner city residential apartment development

Central Park is an urban village located in Chippendale, Sydney, New South Wales (NSW), Australia, on the former old Kent Brewery site. The $2 billion mixed use development comprises two 5 Star Green Star-rated commercial and residential towers. The SATEC EM133AR meter [11] was chosen by the developer as the best fit for the project. The requirements were for a DIN-mounted Class 0.5 s, according to AS/IEC 62053-22, direct connect meter to be installed within each apartment for the monitoring of multiple electrical sources including hot/cold water usage [12].

Central Park operates a 'tri-generation system' and a membrane bio-reactor for the advanced water treatment (recycled water) system to support self-sufficiency and sustainable energy and water consumption. Therefore, fan coil units (FCU) are used instead of individual air-conditioning systems for each residential unit. The objective of measuring cold water and hot water is to allocate the heating/chilling cost to each tenant based on their consumption. The EM133AR was deployed to measure 3x single-phase supplies in each apartment, as well as the cold/hot water. Overall, each meter supports 7x sub-meters per apartment (**Figure 9**):

1. Total electricity;
2. Light + power;
3. FCU #1;
4. FCU #2;
5. FCU #1 + FCU #2;
6. Cold water; and
7. Hot water.

An example of an alternative raw data configuration inputs to the meter and configuration to support the 7x sub-meters per apartment is represented below:

1. Phase 1—light and power (kWh);
2. Phase 2—fan coil unit #1 (kWh);
3. Phase 3—fan coil unit #2 (kWh);
4. Phase 2 + 3—fan coil unit #1 + #2 (kWh);
5. Digital input 1—cold water (kL);
6. Digital input 2—hot water (kL);

7. Digital input 3—recycled or other type water (kL); and
8. Digital input 4—gas meter (m^3).

Figure 9. Meters installed at Central Park: (a) typical complete installation of meters for gas, electricity, hot water, cold water, (b) close up view of the gas and water meters, (c) showing electricity meter EM133AR installed in residential DB with Ethernet output to building IP backbone.

SATEC developed an algorithm specifically for the Central Park project based on the Green Star principles for daily/weekly/monthly reporting. SATEC also added 'quarterly' so that the EM133AR profiles for 'electricity and water' can easily be displayed as 'daily/weekly/monthly/quarterly' and accessible at the meter and at the Modbus/TCP/IP interface levels. The meter was programmed to provide the information over 3× periods, thus available on the meter is profiled:

- Today, yesterday and day prior to yesterday;
- This week, last week and week prior to last week;
- This month, last month and month prior to last month; and
- This quarter, last quarter and quarter prior to last quarter.

The methodology of the in-depth meter program above was designed to comply with the Green Star's Metering and Monitoring Credit. The credit specifies that the Green Star requirements are in line with the CIBSE TM39 Building Energy Metering [4] for best practice in the design of energy metering and sub -metering. Utility meters must meet metering guidelines under the weights and measures legislation, as outlined under the current national measurement regulations.

Non-utility meters (including sub-meters) must follow the same requirements to those described in the most current validating non-utility meters for NABERS ratings protocol, issued by the NSW Office of Environment and Heritage.

The Green Star credit requires all residential premises and tenancies to have their own utility grade meter and access to the meter reliant on accuracy standards. However, points are not awarded to the project without an automated monitoring system installed, which, as a minimum, must be capable of (Green *Star—Design & As Built v1.1 06 Metering and Monitoring*) the below:

http://dx.doi.org/10.5772/64082

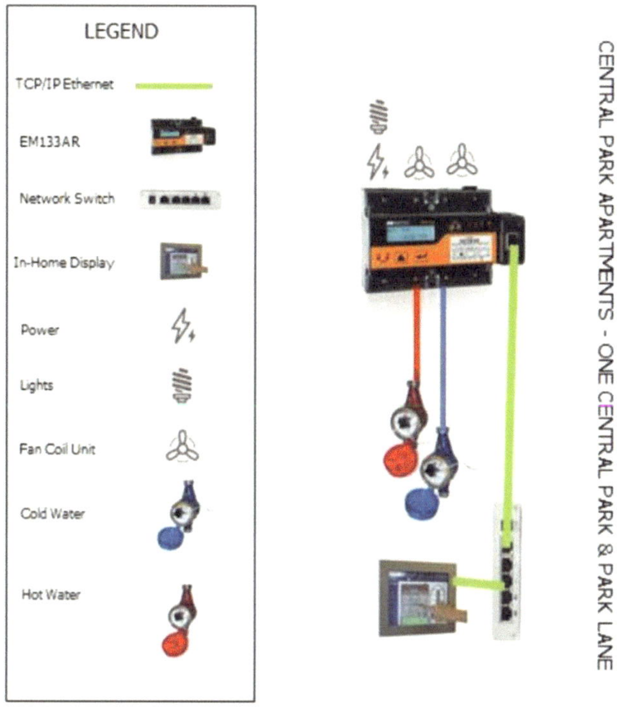

Figure 10. One Central Park & Park Lane design topology.

- Collecting data from all meters;
- Alerting to missing data due to failures;
- Recording and processing of data on energy use or water consumption at user adjustable intervals;
- Raising an alarm when the energy or water use increases beyond certain parameters and automatically and instantly issue an alert the facilities manager;
- Providing a breakdown of the information by building system (mechanical, electrical, etc.), or by space (or by tenanted floor);
- Including the consumption water or energy, the load versus time (load profile), and the power factor (in the case of energy); and
- Producing, as a minimum, a quarterly report that is automatically emailed to the facilities manager responsible for the building.

The SATEC program outlined above provides a virtual utilisation type technique or a distributed logic approach reducing processing power from software platforms particularly for the daily/ weekly/ monthly/ quarterly element. While the program attributes in the meter offer these convenient summaries, real-time data are available through the Modbus register map along with asset information assisting in managing metering assets according to ISO 55000 (**Figures 10–12**).

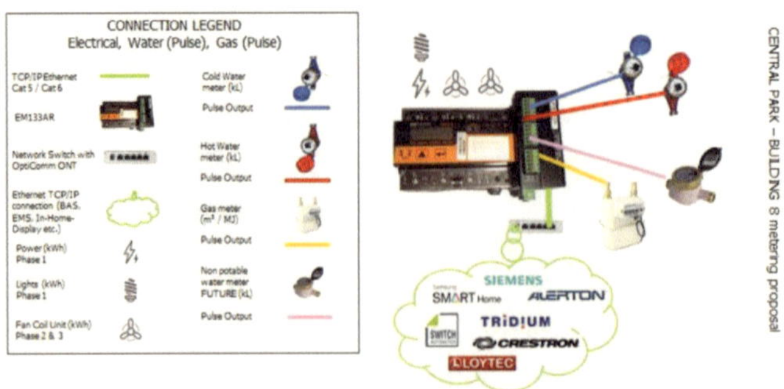

Figure 11. Central Park—Proposed Building 8 Solutions (Mechanical, Electrical, Hydraulic Services).

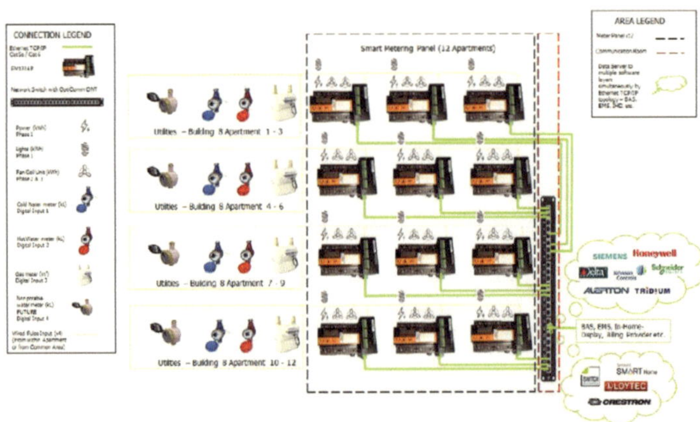

Figure 12. SATEC proposed typical typology for Smart Metering for Central Park Sydney (SATEC EM133AR 'Smart Metering' for Central Park Building 8 Rev 2 04).

The purpose of the embedded network is to allow the developer/owner to have increased financial revenue from their infrastructure. This trend is becoming more popular, as developments and owners are becoming more aware of the benefits that can be gained by owning and operating their utility networks internal to their facilities. This is particularly the case for multi-unit residential developments. SATEC (Australia) Pty Ltd advised that as of December 2015 the embedded network is still not fully operational at Central Park, as the billing provider was not fully utilising the functionality. SATEC (Australia) Pty Ltd further noted that it was possible for them to fully utilise the system, but there were some managerial issues still to overcome.

Common area metering is now more used widely utilised in commercial buildings, mainly due to NABERS and Green Star. Once NABERS became mandatory under the commercial

http://dx.doi.org/10.5772/64082

building disclosure legislation at the point of sale, lease or sub-lease for office space greater than 2000 m^2, metering sales and installations increased dramatically according to SATEC (Australia) Pty Ltd. Comparing to New Zealand, where NABERS is voluntary, fewer owners install embedded network metering systems than Australia. The CBD legislation is under review by the federal government with the recommendation that the threshold of 2000 m^2 be reduced to 1000 m^2, which will further drive the installation of these embedded meter networks from A grade buildings into B and C grades building stock within Australia. The Green Building Council has linked NABERS and Green Star ratings directly to the improvement in return on assets calculations, as specifically expressed in a meeting in Christchurch.

While NABERS energy and water assessments require 12 months of utility meter readings for the building for a whole building reading, a base building rating requires 12 months of tenant meter data to subtract from the buildings total consumption. The outcome is the base building consumption, which is the standardised based upon the climate using the postcode, net lettable area (NLA) and occupancy data to provide the final NABERS energy base building rating.

Alternatively, Green Star requires under Section 6.0.1 Metering Distinct Uses or Floors of its Design & As Built v1.1 rating tool that metering shall be provided to allow for monitoring of the relevant areas or functions of the project. Stating that in most cases floor-by-floor metering will suffice if the entire floor has a single use and that if a floor has multiple uses, the different uses shall be metered. Therefore, should a floor be composed of office space and a seminar room, both spaces shall be separately sub-metered. If a floor has multiple tenants or owners, each tenancy or property shall also be separately sub-metered.

Green Star also has specific requirements where an energy load for a single item exceeds 5% of the total energy use for the building, or 100 kW, it must be independently metered. Supplementary equipment can also be installed on the same measured circuit as the major use item. However, the total combined energy use of any systems connected to the major use item must not contribute more than 10 kVA to the overall energy use.

Examples of systems that are considered to be common uses for energy are provided by the Green Star rating tool, but are not limited to:

- Chillers;
- Air handling units, fans and humidification;
- Server and computer equipment;
- Water reuse systems;
- Kitchen plant and equipment;
- Specialist lighting for stages, etc.; and
- Specialist equipment.

Embedded networks within Australian buildings will gradually change, with the market competition reviews being conducted by the Australian Energy Market Commission (AEMC) and the Western Australia (WA) stand-alone grid known as the South West Interconnected

System (SWIS) being harmonised into the national management framework under changes being made out of the WA Electricity Market Review in particular.

The outcomes of that will be increased competition in metering to be more likely, and advanced metering will be used instead of low-cost metering. The ability for consumers to 'opt-out' of the embedded network, or meter churn, will be a big issue in the future. One way of overcoming this is to leverage improved communication such as Ethernet (TCP/IP). Ethernet is a 'multi-master' communications platform and widely used throughout many industries.

Metrology measuring devices have improved from RS232 communications, which are an example of 1:1 relationships. The communications of RS232 are most demonstrated in networks whereby remote communications is leveraged by modems supporting—2/3G, 4G, GPRS, etc. Metering encompasses further complicated communication networks such as RS485 (e.g. daisy-chain) topology configurations. In particular, Modbus RTU protocol delivered over RS485 methods could limit physical connections with respect to individual modern digital metering systems.

Methods over TCP/IP leverage improved communication methods. For example, a typical 'modem/meter relationship' is limited to a 1:32 ratio in respect to a Modbus RTU protocol implementation. Ethernet leverages modern improvements in communication methods allowing for improved transparent communication methods. The National Broadband Network (NBN) in Australia will inhibit a transition to a modern information enabled generation.

At Central Park in Sydney, Ethernet is used on all meters. The meter has data logging and event logging, as well as a daily/weekly/monthly/quarterly profile. Developed for Green Star reporting principles, the benefit is now the meter is 'the data server'. It performs the 'distributed logic' and alleviates the burden from any given control system such as the building management system (BMS) or building management control system (BMCS). It allows for more efficient communications and works to guarantee that all profiled data are done by the physics of the real-time clock, compared to a traditional 'polling system'.

A traditional 'polling system' is unable to assemble all data correctly across these many meters. Central Park in Sydney has more than 1500 m or devices installed in the digital metering network. Each meter has 7×7× sub-meters per apartment which equates to 10,500 sub-meters on a digital network. The main benefit of Ethernet is the multi-master support. The SATEC TCP/IP module has four TCP sockets available for different software systems to consume the data from the meters simultaneously, which cannot be done with traditional RS485/RS232 communication topologies.

In the case of 'meter churn', or a consumer replacing their respective meter used for trade purposes such as 'billing', future communication developments will enhance the overall consumer experience respective to regional Internet capabilities. Overall 'meter churn' within the Australian jurisdiction, or others, can be improved exponentially, and with effective methods , the overall efficiencies and flexibility in network design improve the following:

- BMS, BMCS, and EMS;

- Billing system;
- In-home display, and;
- Metering programming software (E.g. Power Analysis Software—PAS).

In the future, should a consumer choose to opt-out of the embedded network, an energy retailer could easily access the data directly from the meter through use of TCP/IP Ethernet communications. If ADSL, ADSL+, GPRS, NBN or 2/3/4G are required, then it is with TCP/IP communications whereby enhancements can be made. A simple 'port forward' is managed though IT layers proving a clear and transparent visualisation of the respective metering system. With the use of 'multiple TCP socket', a true disciplined 'multi-master application' can be achieved and overall data limitation through traditional methods can be realised.

7. Conclusions

Smart metering at both the residential and large facility scale should be robust and commissioned correctly. Utilities and urban developers need to engage with communities and consumers to better understand how smart meters and data feedback can best achieve societal goals [13]. This will then ensure accurate and live consumption information is easily accessible to the consumers and utilities where appropriate to enable the most efficient consumption, generation and storage of electricity, water and gas.

These goals can be achieved by choosing meters with standardised high-level communication inputs and outputs, which provide connectivity for additional sub-metered services such as water and gas and ultimately connectivity to a reporting system to the consumer. While the connectivity ultimately should be TCP/IP, for internal university campuses, utilising Modbus and open-source master gateway devices has proven to be a cost-effective structure. Where large-scale multi-tenanted residential or commercial tenancies involved, direct TCP/IP connectivity to the meter provides for future proofing within the Australian market due review of all aspects of the energy landscape and particularly how metering can be leveraged by both consumers and in particular networks and generators.

Energy management systems provide both residential and large energy consumers with the ability to capture, monitor and control their energy consumption and therefore their expenditure whether this is by reducing consumption directly through energy efficiency upgrades or behaviour change to optimise consumption against price signals from the utilities. Accuracy of data is a key element in order to represent the measured outcome for analysis.

Energy management systems combined with structured metering also enable consumers with renewable energy generation such as photovoltaic (PV) panels to monitor their own generation, consumption, import and export. As battery storage becomes integrated with renewable energy generation, consumers will have the ability to consume cheaper renewable energy than can be bought from the grid and sell energy back to the grid at the most economically viable times. While uncertainty surrounds the grid and its impact on rising electricity prices, smart metering, intelligent control systems and utilities offering consumers more amenity and the ability for consumers to participate in the wholesale market will ensure the smart grid can contribute to future carbon neutral [14] urban environments.

Acknowledgements

Material on the Central Park case study was kindly supplied by Dean Davis, SATEC (Australia) Pty Ltd, P.O. Box 82, Mulgoa, New South Wales 2745, dean@satec-global.com.au I www.satec-global.com.au, Tel. +61 2 4774 2959 I Fax. +61 2 4774 0249, Mobile. (Australia) +61 431 421 716.

Author details

Andrew Haning and Martin Anda*

*Address all correspondence to: m.anda@murdoch.edu.au

Academic Chair of Environmental Engineering School of Engineering and Information Technology Building 245, Room 3.017 Murdoch University, Western Australia

References

[1] Sharma K. & Saini L.M. (2015), *Performance analysis of smart metering for smart grid: An overview.* Renewable and Sustainable Energy Reviews, 49, , pp. 720–735, Elsevier.

[2] Anda M., Brueckner M. & Haigh Y. (2016), *Transitions to a post-carbon society: scenarios for Western Australia,* Chapter 15 in "Renewable Energy in the Service of Mankind" Vol II [ed. Sayigh, Ali - PO Box 362, Brighton BN2 1YH, UK], Springer.

[3] Anda M. & Temmen J. (2014), *Smart metering for residential energy efficiency: The use of community based social marketing for behavioural change and smart grid introduction.* Renewable Energy, 67, pp. 119–127.

[4] CIBSE (2009). *TM39 Building Energy Metering (a guide to energy sub-metering in non-domestic buildings).* ISBN 9781906846114. London.

[5] Goodfield D., Anda M. & Ho G. (2014), *Carbon neutral mine site villages: Myth or reality?* Renewable Energy, 66, pp. 62–68.

[6] Anda M., Le Gay Brereton F., Brennan J. & Paskett E. (2013), Smart metering infrastructure for residential water efficiency: Results of a trial in a behavioural change program in Perth, Western Australia. In: *Proceedings of the First International Conference*

on Information and Communication Technologies for Sustainability, 14–16 February, Zurich, Switzerland.

[7] Cominola A., Giuliani M., Piga D., Castelletti A., Rizzoli A.E. (2015), *Benefits and challenges of using smart meters for advancing residential water demand modeling and management: A review*, Environmental Modelling & Software, 72, , pp. 198–214

[8] Xu P et al. (2015), *Case study of smart meter and in-home display for residential behavior change in Shanghai, China*. Energy Procedia, 75, pp. 2694–2699

[9] Cominola A., Giuliani D., Piga D., Castelletti A., Rizzoli A.E. & Anda M. (2015), *Modelling residential water consumers' behaviours by feature selection and feature weighting*. E-proceedings of the 36th IAHR World Congress, 28 June–3 July, 2015, The Hague, the Netherlands. http://89.31.100.18/~iahrpapers/83100.pdf Downloaded 25 February, 2016.

[10] LeeK., Murray D., Goodfield D. & Anda M. (2012), Experiences and issues for environmental science sensor network deployments. In: The 6th IEEE International Conference on Digital Ecosystem Technologies—Complex Environment Engineering, Campione d'Italia, Italy, 18th–20th June, 2012.

[11] SATEC EM133AR TOU Smart Energy Meter for Central Park Building 8 Rev 2 04. Sydney, Australia.

[12] SATEC Australia (2014). *One Central Park by SATEC Australia*, Scoop Publishing, August 2014. http://www.scoop.com.au/Online-Articles/One-Central-Park-by-SATEC-Australia. Accessed 18 Feb 2016.

[13] Bertoldo R., Poumadère M. & Rodrigues L.C. (2015), *When meters start to talk: The public's encounter with smart meters in France*. Energy Research & Social Science, 9, pp. 146–156.

[14] Beattie C., Bunning J., Stewart J., Newman P. & Anda M. (2012), *Measuring carbon for urban development planning*. The International Journal of Climate Change: Impacts and Responses, 3(4), pp. 35–52.

Chapter 5

MAC Protocol Design for Smart Meter Network

Yue Yang

Additional information is available at the end of the chapter

http://dx.doi.org/10.5772/62392

Abstract

The new generation of power metering system—that is advanced metering infrastructure (AMI)—is expected to enable remote reading, control, demand response and other advanced functions, based on the integration of a new two-way communication network, which will be referred as Smart Meter Network (SMN). In this chapter, we focus on the design principles of multiple access control (MAC) protocols for SMN. First, we list several features of SMN relevant to the design choice of the MAC protocols. Next, we introduce some performance evaluation metrics and give a survey of the associated research issues for the SMN MAC protocols' design. In addition, we also note progress within the new IEEE standardization task group (IEEE 802.11ah TG) currently working to create SMN standards. After that, in order to emphasize the importance of the performance metrics mentioned before, we give several MAC protocol design examples which could solve the associated research issues and challenges for the SMN.

Keywords: Smart Meter, MAC protocol, AMI, Smart Metering Network, Grouping

1. Introduction

Since the Smart Meters application crucially depends on two-way networks, denoted as Smart Meter Network in this chapter, communication aspects of Smart Meter Network design have begun to draw attention [1–4]. Some of those discuss the choice of communication architectures [5] that are appropriate for the various AMI applications, so as to achieve the traditional goals of communication reliability, efficiency and security. On the other hand, some papers talk about the pros and cons of multiple different communication technologies, such as power line communications [6], ZigBee [7] and WiFi [8], and compare their suitability to the AMI applications. Furthermore, one new IEEE 802 standardization task group (802.11ah TG) [9] has been established and aims to create new standards to provide a guideline for the design of Smart Meter Network. In addition, as an important part of network design issue, the multiple access control (MAC) protocols' design, which is used to regulate the data transmission on the shared channel and largely determines the efficiency of end-to-end data collection process, starts to become the research focus recently. Therefore, in this chapter, we first discuss some SMN unique features that will significantly impact the choice of MAC protocols in Section 2. After that, we discuss several MAC protocols' performance metrics and design challenges especially in Smart Meters Network in Section 3. In Section 4, the progress of the IEEE 802 standardization task group (802.11ah TG) in the aspect of MAC protocols' design is also reviewed. After that, we present several solutions and examples of the MAC protocols' design, presented in [10, 11], which adapt to the special requirements of Smart Meter Network and solve the challenges mentioned below.

2. Smart Meter Network features

There are several features in Smart Meter Network that have significant impact on the MAC protocol design. These features are listed in **Table 1**, and two of them are introduced in more details in the following subsections.

Important features	Brief discussions
Smart Meter Network architecture	• Smart Meter <-> Local Collector <-> Central Collector • MAC protocols' design mainly focuses on Neighbourhood Area Network (NAN), that is Smart Meter <-> Local Collector. • Many Smart Meters are associated with one Local Collector
Candidate communication technologies	• Power line communications • ZigBee • Machine-type communications in cellular communications • WiFi-based communications
Fixed location	• Smart Meters are implemented in fixed location without mobility concern in MAC protocol design.
Information redundancy	• Smart Meters information has nearly zero redundancy but may be highly correlated. This feature should be considered in MAC protocol design with respect to accurate data aggregation or information estimation.

Table 1. Smart Meter Network features outline.

http://dx.doi.org/10.5772/62392

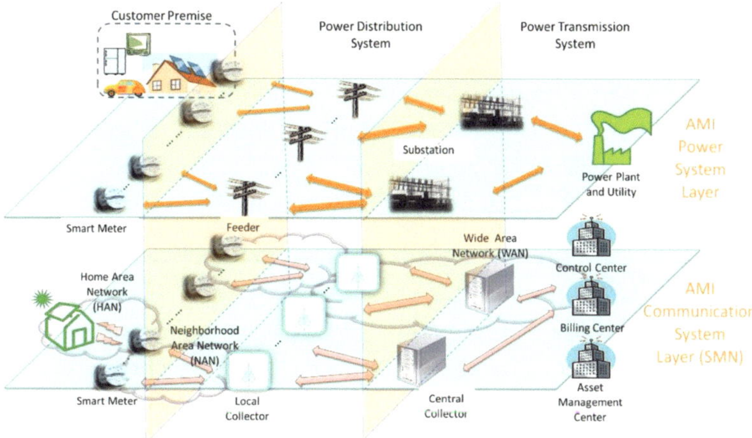

Figure 1. Smart Meter Network system architecture.

2.1. Smart Meter Network system architecture

Figure 1 presents an SMN system based on the Smart Grid architecture given in [12]. A short overview of its main components is as follows:

- Smart Meter (SM): This device has three different roles. First, the Smart Meter is a multi-utility instrument measuring electric power consumption (and possibly in future, gas, water and heat). It can thus act as an energy control centre, that is as a point of aggregation for usage information collected using a home area network (HAN) that connects home appliances. Finally, the Smart Meter also serves as the gateway between HAN and external network; it reports on energy consumption, sends out urgent data, receives remote commands from the utility and is responsible for security of the above transactions. It is noted that SMs' nodes are fixed as they are deployed in customer premises. They are usually powered from the main supply, and hence, power-saving issue is not as important as in traditional battery-powered wireless sensor network nodes.

- Home Area Network (HAN): This is composed of multiple interconnected electric appliances, such as air conditioner, dish washer, plug-in hybrid electric vehicles (PHEV), and the Smart Meter. All the components inside HAN share information or deliver control commands to each other. For example, the dish washer may send a signal to Smart Meter, requesting it to send a 'postpone' command to the charging PHEV, so that it may operate without incurring excessive energy cost at that time.

- Local Collector (LC): Between the Smart Meter and Utility Centre, there could be multiple layers of intelligent electronic devices (IEDs) that act as data concentrators. For example, a data collection node closer to customer premises—named as Local Collector—collects SM data from multiple premises and relays it to the Central Collector. Additional functions at the Local Collector may include simple data processing and distributed decision and intelligence using its own data. The network segment between Smart Meter and Local

Collector is called neighbourhood area network (NAN), while that above Local Collector belongs to wide area network (WAN).

- Central Collector: A centralized data repository for the entire region operated by the utility that acts as the interface with Control Centre, Billing Centre and Asset Management Centre. These centres may use this data to conduct analysis and evaluate system status, and make decisions or deliver control commands to other components.

The mapping between the SMN components and their prospective physical deployments is also shown in **Figure 1**. For example, the Local Collector could be located at a distribution transformer because it would be easy to power the Collector and obtain measurements from other feeder devices. On the other hand, the independent deployment for Local Collector may provide more flexibility to adjust its coverage range. The Central Collector is likely to be deployed closer to the centres. If the utility's coverage region is not very large, only one Central Collector located close to the centres may suffice. Otherwise, Central Collectors may be placed at the distribution substation as the second tier data relay. Since the HAN consists of electric appliances which are manufactured by different vendors and has much flexibility in implementation especially on application layers, the utility operating SMN may leave it open and focus their design on the upper level network. On the other hand, the design of network from the Local Collector to the Central System does not only depend on the communication requirements of SMN because it also includes the electrical devices which serve the power systems other than AMI. Therefore, the main focus of the MAC protocols' design for SMN lies on the segment from Smart Meter to Local Collector (NAN), which is exactly also the focus in this chapter.

2.2. Candidate communication technologies

A large amount of literature discusses the feasibility of several optional bi-directional communication technologies applied on the SMN. We summarize the various options and highlight their pros/cons in **Figure 2**.

Technology	Applications	Benefits	Limitations
Power Line Carrier	WAN, NAN, HAN	No Extra Cabling Fee, High Security	High Noisy Medium, Low Scalability
Messaging over Cellular Network	HAN, NAN, WAN	Mature Development, Long Range	Low Data Rate, Low Robustness, Low Security, Costly Spectrum Fees, Low Scalability
WiFi	HAN, NAN, WAN (with multi-hop)	Mature Development, Free License, High Robustness	Low Security, Low Scalability
ZigBee	HAN	Low Cost	Short Range, Low Security, Low Data Rate

Figure 2. Comparison among optional communication technologies for SMN.

In power line carrier (PLC), the data are transmitted over electricity transmission lines along with electrical power [13]. Its communication performance depends on several factors, such as frequency, propagation distance and existence of transformers because the data signals cannot go through the transformers. PLC has gained a lot of attractions because it uses the existing power lines as signal carrier and no extra cabling fee is needed. Therefore, many countries (e.g. Singapore) adopted it for broadband communication services. However, PLC also suffers from several disadvantages, such as high-signal attenuation, high noisy medium and lower scalability, which lead to the termination of deployment in some countries (e.g. US) [14].

ZigBee is a wireless communication technology that consumes low power at the device side [13]. Thanks to its low cost and easy implementation, this technology has already been widely used in the Smart Home Network by many AMI vendors, such as Itron and Landis Gyr. They produce Smart Meters and measuring devices integrated with ZigBee protocol to monitor and control the Home Energy Status. On the other hand, there are still some constraints on ZigBee for its practical application on the SMN. For example, its short range confines this protocol in the application domain of HAN. Furthermore, the processing capabilities and memory size of the ZigBee device are expected to be improved for more advanced functions and communication requirements of the SMN.

Machine-type communications over cellular network allows the Smart Meters and Local Collector to exchange information via low data load communication service, which has been supported by multiple mature cellular network standards, such as LTE [15]. It is the popularity and easy implementation that make this technology become an attractive candidate option. Furthermore, the long range and high data rate provide the utility more flexibility to design and implement the SMN. However, the concern about reliability, security and delay performance makes a barrier for the implementation of this technology in practice, especially under the condition of heavy traffic load.

Finally, WiFi is a communication technology that allows devices to exchange data wirelessly based on IEEE 802.11 Standards. Its popularity, mature development and unlicensed spectrum make it on the top of the candidate technology list. Furthermore, it is also a cost-efficient network with dynamic self-healing and distributed control, which makes it easier to be implemented. On the contrary, the capacity, scalability and security issues are the main challenges for its application on the SMN. Therefore, in order to solve these challenges, a new standardization task group IEEE 802.11ah is established and aimed at creating a WiFi-based standard to support wireless communication between Smart Meters and Local Collector as one of its primary use cases. In the following sections, we also regard this WiFi-based communication technology as our main foundation of MAC protocols' design.

3. MAC protocols design performance metrics and important research issues

MAC protocols must be designed to match the different objectives for the various types of Smart Meter data as well as adapt to the different network topology scenarios. Furthermore, the special features and applications in Smart Meter Network also address some new challenges to the suitable MAC protocols' design. In this section, we outline the broad performance metrics for MAC protocols as they relate to Smart Grid operations and identify some specific challenges for MAC protocols' design in Smart Meter Network.

3.1. Different data types

Usually, the data transmitted over Smart Meter Network may be classified according to the latency requirements. For example, energy consumption information is delay tolerant compared to fault reports or other control actions in response to some urgent events that must be communicated as soon as possible [16]. On the other hand, the Smart Meter traffic may also belong to two different classes: periodical and event-triggered data. The former includes energy consumption information, while the latter is largely data from protection devices (relays, reclosers, etc., that monitor local fault status) or electric vehicle charging stations. We list a table of several representative traffic examples with their important properties in **Figure 3**.

Traffic Examples	Delay Requirements	Trigger Type
Outage Alert	Seconds	Event Triggered
Billing Information	Minutes to Hours	Periodical
Demand Response	Seconds or Minutes	Periodical
Real-time Pricing Information	Seconds or Minutes	Periodical
EV Charging Information	Seconds or Minutes	Event Triggered

Figure 3. Examples of data with different communication requirements.

3.2. Scalability

Scalability requires that MAC protocols continue to perform well as the number of Smart Meters offering data scales, which is the top challenge to MAC protocols' design in Smart Meter Network. Referring to [17], one Local Collector is required to support a network associated with a large number of Smart Meters, which are much higher than that covered in the traditional WLAN network, such as WiFi. For example, as specified in the Use Case 1a: Smart Grid—Meter to Pole requirement in [17], the capacity of one Local Collector (AP) is required to be upto 6000, which is usually the network scenario in densely populated urban city in Asia. According to [18], the urban outdoor path loss model for 900 MHz RF in dB is as follows

$$P_{L,dB}(r) = 8 + 37.6 \times \log_{10} r, \qquad (1)$$

where r denotes the distance between the Smart Meter and Local Collector. Then based on the parameters listed in the table (**Figure 4**) [9], the received power and noise power at the receiver terminal are given as follows:

$$P_{RX,dB}(r) = P_{TX,dB} + G_{dB} - P_{L,dB}(r) = 0 + 3 - (8 + 37.6 \times \log_{10} r) \quad (2)$$

$$P_N = k \times T \times W = 1.3 \times 10^{-23} J / K \times 290K \times 2MHz \quad (3)$$

Referring to the AWGN capacity derivation with BPSK modulation in [19], we may draw a figure of the transmission rate and received power with respect to r. As shown in **Figure 5**, the RX received power at 1200 m satisfies a reasonable received power threshold (−120 dB) and achievable transmission rate at 4500 m exceeds 100 kbps, the minimal required data rate set in IEEE 802.11ah Use Case [17]. This implies that one Local Collector may need to communicate with individual Smart Meters located 1200 m away using a star topology. Given typical SM density ρ (1000–6800 SMs per km^2) [20, 21], it is possible to have in excess of 6000 Smart Meters communicating to one Local Collector. Clearly, the design of MAC protocols for such large number of nodes invites challenges about scalability of any chosen MAC protocol for SMN as discussed next.

Parameter	Value	Parameter	Value
Channel Bandwidth W	2 MHz	Transmitter Power $P_{TX,dB}$	30 dBm
Antenna Gain G_{dB}	3 dB	Temperature T	290 K
Transmission Frequency	900 MHz	Boltzmann's Constant k	1.38×10^{-23} J/K
SM Density ρ	1000 per km^2	Cell Radius R	1200 m
$SINR_{dB,th}$	14 dB	$P_{RX,dB,th}$	-129 dB

Figure 4. Parameter list for transmission rate and hidden node calculation.

Taking random access protocols based on carrier sensing as an example—such as DCF (CSMA/CA) in IEEE 802.11—such a large coverage area corresponding to a single Collector cell will lead to significant hidden nodes. In the traditional random access protocols, such as CSMA/CA, all nodes listen to estimate channel status (busy/idle) based on energy threshold. If the node observes the channel to be continuously idle for a specific interval, it starts contending for the channel via a random back-off process.

However, when the coverage of the Local Collector is enlarged, one Smart Meter (hidden node) may not be able to detect ongoing transmission of other meters due to the degenerated radio channel condition. Then, this hidden node may initiate its own transmission because it determines the channel as idle, which leads to a collision.

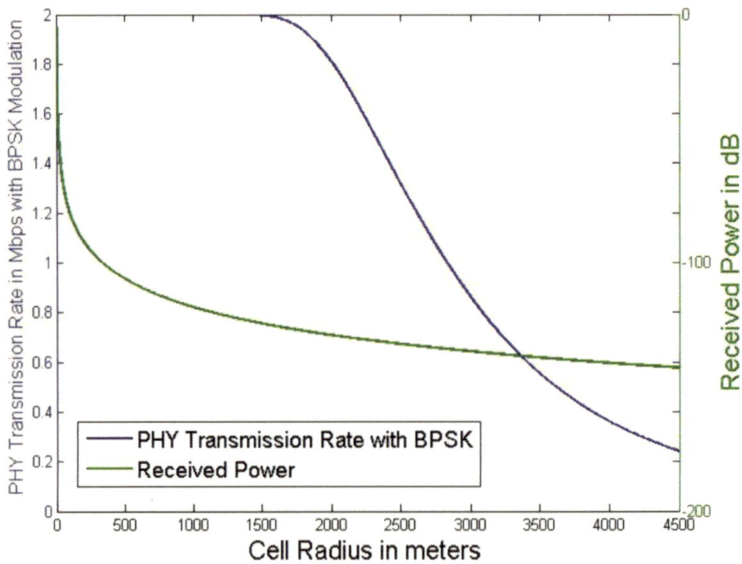

Figure 5. PHY transmission rate of the communication with BPSK modulation and received power at Local Collector with respect to their distance.

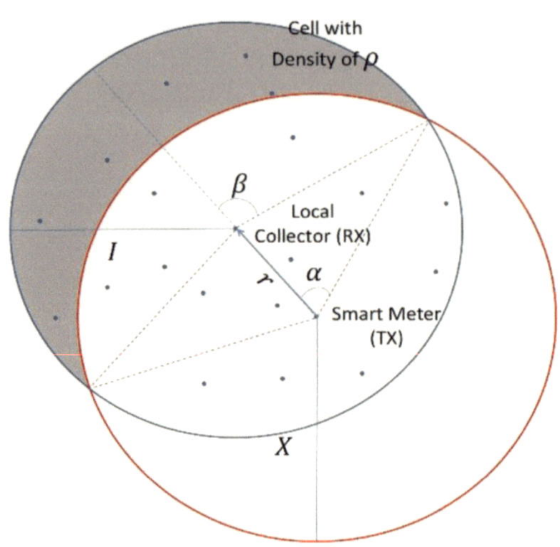

Figure 6. Network topology for hidden node calculation.

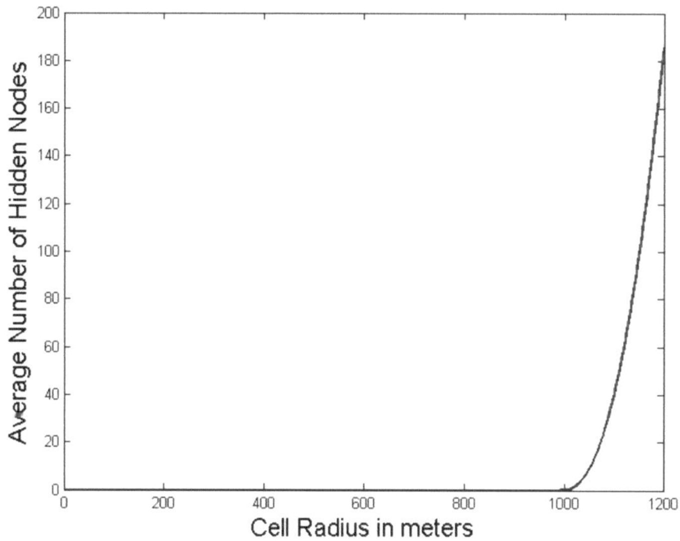

Figure 7. Results for hidden node calculation.

In order to investigate the number of hidden nodes with growing Local Collector coverage, we consider the network topology of a disk with radius of R and uniformly distributed SM density ρ (**Figure 6**) and the distribution of Smart Meter deployment with respect to r is $f(r) = 2r/R^2$, where $0 \leq r \leq R$. Since the down-link communication from Local Collector may cover all the Smart Meters successfully, thus the hidden node only exits in up-link communications where Smart Meter is always the transmitter (TX) and Local Collector is always the receiver (RX). According to [22], the area inside the interference range of the receiver I and outside the carrier sensing range of the transmitter X is defined as the hidden area $A(r)$ (shadowed zone) for a given Smart Meter, which is given as follows:

$$\textit{If } r \leq X - I, \; A(r) = 0$$

$$\textit{If } r > X - I, \qquad A(r) = \beta I^2 + rX\left|\sin(\alpha)\right| - \alpha X^2 \tag{4}$$

where $\alpha = \cos^{-1}(X^2 + r^2 - I^2/2rX)$ and $\beta = \pi - \cos^{-1}(I^2 + r^2 - X^2/2rI)$.

The carrier sensing range X can be calculated based on the equation $P_{TX,dB} + G_{dB} - P_{L,dB}(X) = P_{X,dB,th}$, where $P_{X,dB,th}$ is the carrier sensing threshold for the received power such that the received signal can be detected. Furthermore, the interference range I can be derived based on the following equation:

$$I(r) = \min\left[R, \left\{y | P_{RX,dB}(r) - 10\log_{10}\left(P_N + P_{RX}(y)\right) = SINR_{dB,th}\right\}\right] \tag{5}$$

where $SINR_{dB,th}$ is the threshold for signal noise interference ratio such that received signal can be decoded successfully.

After that, we use $N_{hidden} = \int_0^R \rho A(r) f(r) dr$ to obtain the mean number of hidden nodes inside the network and plot it with respect to cell radius R based on the parameters listed in **Figure 4**. As shown in **Figure 7**, the number of hidden nodes increases dramatically when the coverage of the network is enlarged.

Although the DCF in 802.11 proposes the RTS/CTS algorithm to reduce the occurrence probability of hidden nodes event, its effect on such a large area network still needs to be investigated. On the other hand, the increasing number of Smart Meters may also increase the data load in the Smart Meter Network, which basically results in stronger competition for the medium access. Then, the collisions and following retransmissions happen more frequently, which directly aggravate the performance. Therefore, whether the MAC protocols may guarantee the latency requirements of the event-driven and delay sensitive data under the case of large network is also a critical metric to evaluate the designed MAC protocols.

3.3. Delay

Different types of data induce different communication requirements within a Smart Meter Network. For example, delay sensitive data—such as those reporting a fault and protection related messages—should have higher priority over others, so as to minimize end-to-end latency. Therefore, how to minimize the Smart Meter Network latency—such as that of the last hop between the Smart Meter and the Local Collector is a primary concern. In general, several factors impact the delay, such as the choice of the communication technology, the network architecture, and most notably, the MAC protocol.

A good MAC protocol can coordinate the uplink transmissions of multiple communication nodes to reduce the collision probability significantly, resulting in lower delay. Usually, it is convenient to design MAC protocols for one type of traffic—for example, random access MAC protocols such as distributed coordination function (DCF) in 802.11 have been designed to provide reasonable efficiencies in terms of throughput at low-to-moderate loads, but the delays escalate rapidly as the average load increases. On the other hand, a polling-based (taking-turns) protocol such as point coordination function (PCF) defined in 802.11 is well-suited to reporting data with bounded delay guarantees. The total duration of one polling cycle increases only linearly with the number of nodes (in contrast to exponential increase in delay with random access systems as the aggregate load increases) and provides a guaranteed delay bounds. However, the efficiency of such protocols declines rapidly with the number of nodes, providing a different trade-off to random access protocols.

However, most Smart Grid scenarios comprise of a mix of traffic, for example regular traffic and emergency traffic. To serve both types within a DCF framework, the notion of traffic classes were introduced via enhanced distribution channel access (EDCA) defined in 802.11e [23], to prioritize low latency data over non-time-critical data applications (such as billing information). A combination of EDCA and PCF, hybrid coordination function controlled channel access (HCCA), tries to serve multiple traffic types by granting higher priority to some particular kinds of data via polling algorithm, which is centralized controlled by access point (AP). On the other hand, the performance against scalability issue of these hybrid MAC protocols under densely populated network still needs to be evaluated.

3.4. Fairness

This seeks to measure whether each node in the Smart Meter Network obtains a fair share of system resources. Fairness can be quantified in terms of the access probability to the shared channel by each node—ideally, this should be equal (independent of the node) assuming that all Smart Meters require identical data rates. In general, the notion of proportional fairness should be applied, based on different data rate requirements by different nodes.

3.5. Security

Data security in Smart Meter Network is an extremely vital issue as it relates to household or customer information (e.g. energy consumption profile) that is considered private.

Therefore, it is necessary to encrypt the message to prevent eavesdroppers from intercepting the message. Although cryptographic tools and algorithms are relatively mature, these will result in extra load on the Smart Meter Network. An open question is whether the known features of Smart Meter data may be exploited to develop simplified yet effective cryptographic approaches. Secondly, end-point authentication is also indispensable for Smart Meter Network; whenever the data collector receives a message of energy consumption report, it has to authenticate the identity of the sender.

Specifically, defences against two common types of attacks will be of high priority. Integrity in data communications between Smart Meter and Local Collector may be compromised by a relay or man-in-middle attacker. And such communications may be targeted for disruption via denial-of-service (DoS) attacks by saturating the Local Collector with a large number of spurious external communications, so that it cannot respond to the legitimate traffic [24]. Within this context, it is noted that most Smart Meter communications are regular as it reporting actions are typically scheduled. Therefore, we may exploit such features to filter out malicious accesses by an attacker, by identifying anomalous access traffic patterns.

On the other hand, a good design of the MAC protocol with the aid of pseudo-random algorithm can also effectively protect the privacy of the customer data.

3.6. Expandability

For Smart Meter Network, the expandability means the ability of this network to accommodate the new communication nodes (Smart Meters) to its existing capacity. It is noted that the deployment of Smart Meters will not occur according to a fixed schedule; for example, whenever a house is built, a newly installed Smart Meter will need to be introduced into the existing NAN covered by the corresponding Local Collector. This introduction procedure, which may include registration, identity authentication, geographical location identification, etc., has to be conducted automatically and is used for the Local Collector to determine the newly installed Smart Meter is ready to work inside its coverage. Therefore, how to realize this introduction procedure should be a part of the MAC protocols' design. For example, whenever one newly installed Smart Meter is online, it sends a request to its associated Local Collector to report its own identification and geographical location. After that, the collector registers this new meter and replies it via a message with some necessary setting information. Then, how to automatically modify the parameters of current communication systems due to the newly registered Smart Meter still needs to be analysed.

3.7. Fault detection

For Smart Meter Network, the fault can be categorized into two kinds of cases. The first one is data fault, which means the data involved in the message has some errors. These data errors may be caused by monitoring errors or malicious message altering. Fortunately, the data collector may detect such kind of fault by some statistical algorithms, such as comparison between the current data and historical data. This kind of fault and corresponding solutions mainly occur at application layer. What the MAC protocols' designers need to consider is another one, communication fault, which means that some Smart Meters cannot communicate with the Local Collector directly. These communication problems may be caused by the malfunction of Smart Meters or the degeneration of wireless communication environment since communication fault leads to the Smart Meter Network as quickly as possible. For example, the Local Collector may exploit the idle communication intervals to poll every Smart Meter and expect its feedback. After that, the collector may detect the silent meters by checking the missing feedbacks.

Furthermore, how to schedule the poll-feedback actions in detail and improve its efficiency still need to be investigated.

4. IEEE 802.11ah standardization task group

The success and popularity of IEEE 802.11 (WiFi) enabled communication devices have led to a new standardization task group IEEE 802.11ah, aimed at creating a WiFi-based standard to support wireless communication between Smart Meters and Local Collector as one of its primary use cases. According to [9], IEEE 802.11ah compliant devices will utilize multi-input, multi-output-orthogonal frequency division multiplexing (MIMO-OFDM) at frequencies below 1 GHz, where there are no licensing and regulatory issues. The most discussed channelization for 802.11ah in the US focuses on the 902–928 MHz band, which is currently free. The 802.11ah Working Group appears to have settled on 1 and 2 MHz as the possible channel bandwidth [25]. Besides the benefit of free spectrum, the signal transmission below 1 GHz generally suffers less propagation path loss, enabling the network to achieve larger coverage, as verified above.

With respect to the MAC protocols' design for the use case of Smart Meter—Local Collector communications—the IEEE 802.11ah TG is considering using improved DCF and PCF [9]. Since the area of a cell for one Local Collector may cover thousands of Smart Meters that far exceed the current capabilities of base DCF protocol that was intended for small, indoor cells serve 20 users on average. Hence, the scalability issue of DCF is one of their main discussion areas. The modified DCF with contention factor and prohibition time is one of the suggestion new options. Before the contention phase, the Local Collector broadcasts a prohibition time T and contention factor $0 < Q < 1$, according to the current network congestion status. After that, each Smart Meter generates a random number r which follows a uniform distribution on the unit interval and compare it to Q. Then, the Smart Meter may contend for the channel if $r > Q$ and, otherwise, it keeps silent until the prohibition time T passes. In order to further relieve contention congestion, the MAC scheme may also divide all the Smart Meters within a cell into several groups and provide different groups with different parameters Q and T or allow them to contend for the channel group by group. The collision probability is expected to decrease dramatically (compared to traditional DCF applied to all Smart Meters), as a result.

In order to solve the scalability challenge for PCF, IEEE 802.11ah TG proposes a modified PCF scheme, Probe and Pull MAC (PP-MAC) [9]. After partitioning all the Smart Meters into groups, the Local Collector broadcasts a probe message to a certain group of Smart Meters before the contention free phase. After that, the Smart Meters having data to send reply a short Probe-ACK concurrently with the use of Zadoff–Chu sequences. By assigning these orthogonal sequences to each Smart Meter and multiplying their own messages with their respective sequences, the cross-correlation of the simultaneous short probe-ACK transmissions is reduced, so that the Local Collector is able to resolve these parallel ACKs and identify the different transmitters. After that, the Collector schedules and only polls the Smart Meters with Probe-ACK, which leads to a shorter polling cycle and a more efficient taking-turns MAC protocol.

5. MAC protocols design example I: grouping-based MAC protocols for EV charging data transmission in Smart Meter Network

As mentioned above, the number of Smart Meters involved in Smart Meter Network (a single Local Collector coverage) is much larger than those in today's local area networks, the traditional random access MAC protocols such as DCF in IEEE 802.11 does not work well on event-triggered data communication, for example EV charging data transmission, due to the scalability issue. Therefore, we propose two grouping-based MAC protocols: TDMA-DCF (TDCF) and Group Leader DCF-TDMA scheme (DCFT) in the paper [11]. These two schemes are directed at 802.11 type networks operating at the frequencies below 1 GHz that has been adopted by a new IEEE 802.11ah standardization task group.

In TDMA-DCF, all Smart Meters are divided into several groups and separating network channel access into two-tiers: inter-group and intra-group tier. During each periodic frame in the contention phase, the Local Collector allocates one mutually exclusive sub-frame to each group by broadcasting a control message.

Once a generic group has been allocated the sub-frame and obtains the transmission rights, all the Smart Meters inside this group compete for the channel by DCF. Therefore, the number of meters involved in the contention concurrently is the number of Smart Meters per group at most. The random access at intra-group tier lasts until the sub-frame duration passes or the channel keeps idle for a continuous interval, defined as 'idle interval'. After that, the Local Collector broadcasts another control packet to allocate a new sub-frame to another group. It is noted that there may exist some communications not finishing at the end of sub-frame, and then it may collide with the random access of next group nodes if they do not detect the ongoing transmission. Therefore, in order to avoid such collision, the sub-frame allocation is only controlled by the Local Collector's broadcasting message. The scheme is operated as shown in **Figure 8**.

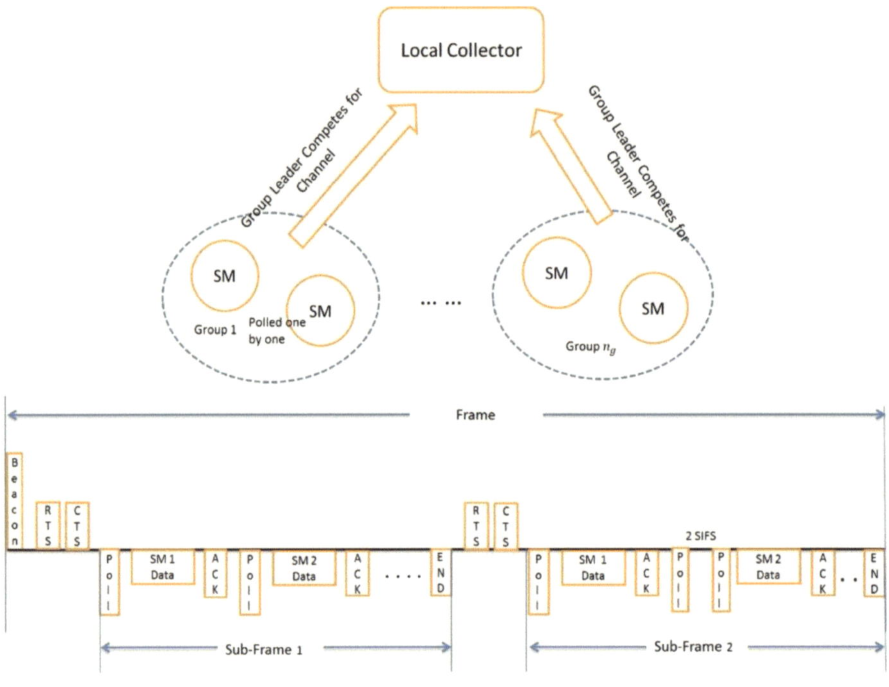

Figure 8. TDMA-DCF scheme operations.

In order to reduce the hidden node events of the random access at intra-group tier, we divide all the Smart Meters into several groups according to their deployment proximity and optimal group size. After the division, as shown in **Figure 9**, the grouping status is recorded in the Registration Table.

On the other hand, in the Group Leader DCF-TDMA scheme, we divide all the Smart Meters into groups according to a different rule which is mentioned below. Furthermore, the Group Leader DCF-TDMA scheme applies DCF at the inter-group tier and Polling protocol at the intra-group tier, which is opposite to the TDMA-DCF scheme. First, Local Collector assigns a group leader for each group. At the beginning of the periodic frame of the contention phase, all the group leaders compete for the channel via RTS/CTS exchange based on the DCF scheme.

http://dx.doi.org/10.5772/62392

Therefore, the number of Smart Meters involved concurrently in the random access at most equals to the number of groups which is much less than the total number of meters. If a generic group leader wins the competition, that is this leader reserves a sub-frame for its associated group via the RTS message and informs other group leaders of the expected duration via the CTS responded by the Local Collector, and then the Local Collector polls all the meters inside the 'winner' group one by one; the Smart Meter replies to the Local Collector with its own data sequentially. The reserved sub-frame lasts until the Local Collector broadcasts an 'END' message when all the meters inside the group finish the transmissions. If any meter in the group has nothing to transmit and keeps silent to the Polling message, the Local Collector is able to detect this case and starts to poll next sequential meter after two continuous idle SIFS. Accordingly, the collector may broadcast the 'END' signal earlier. After that, all the group leaders start to compete for the channel again. It is noted that in this scheme, the Polling operation is centralized controlled by the Local Collector. Therefore, our scheme is more suitable for the use case of SMN, because the group leader in SMN is just the normal Smart Meter with simple infrastructure. The operation of this scheme is outlined in **Figure 9**.

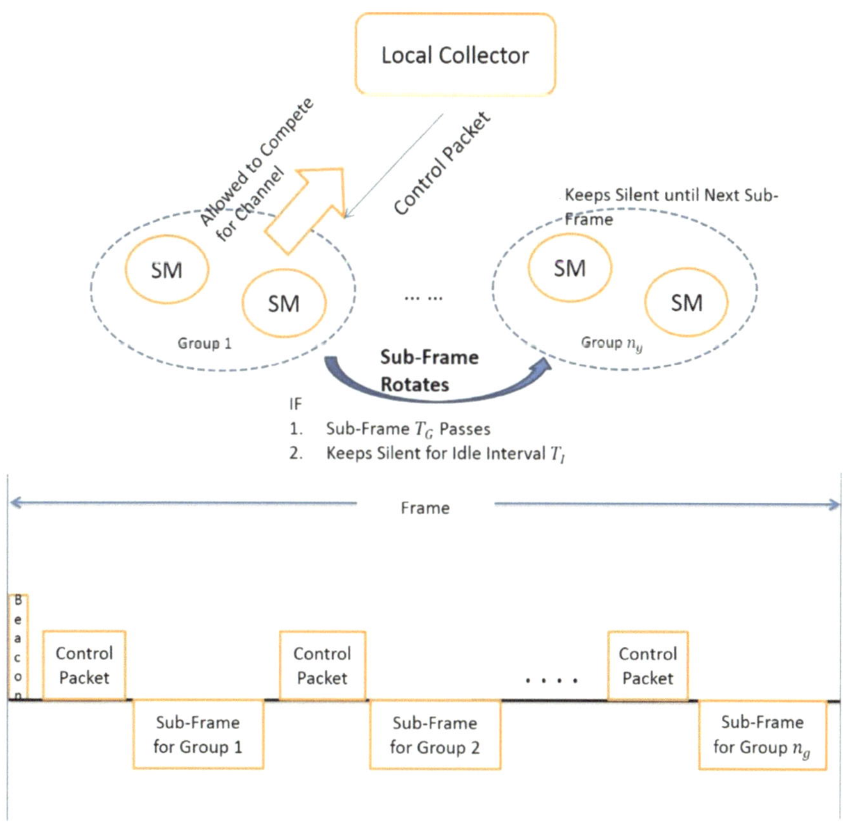

Figure 9. DCF-TDMA scheme operations.

On the contrary to the TDMA-DCF scheme, in order to reduce the hidden node events of the random access at inter-group tier, we have to select the Smart Meters located close to the Local Collector to act as the group leaders, among which there is no hidden node problem. After that, all other Smart Meters may be randomly picked to associate to each group leader and form the group. After division, as shown in **Figure 10**, the grouping status is also recorded in the Registration Table.

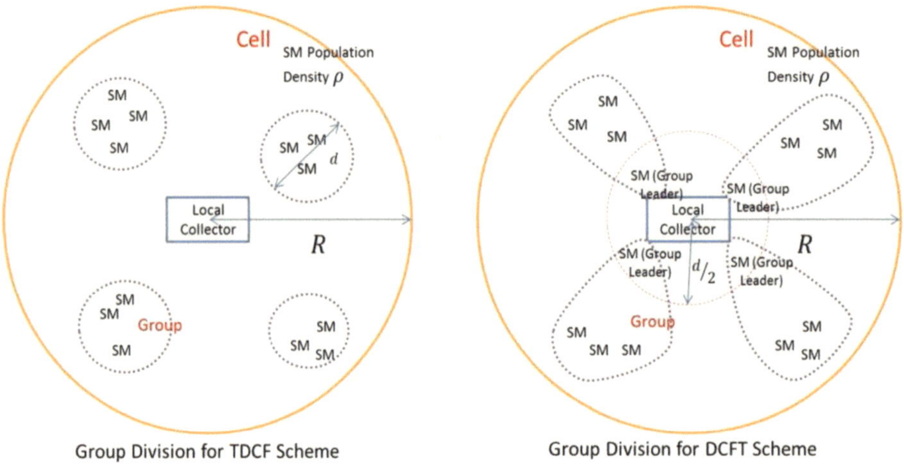

Figure 10. Network topology after group division.

Furthermore, from the calculation in [11], the approximate value of successful transmission distance is 1600 m. If we model it as a disk with radius 800 m, the number of Smart Meters within this area is about 2000. That means, as long as the group size for TDMA-DCF scheme and the number of group leaders for DCF-TDMA scheme is smaller than 2000, it is believable to assume that there are few hidden nodes in these two grouping-based schemes.

The protocol details and comprehensive throughput and delay analysis on these schemes are presented in [11]. The numerical results show that these two grouping-based MAC protocols significantly outperform the traditional random access protocol DCF in a densely populated network with large coverage like Smart Meter Network.

6. MAC protocols design example II: enhanced PCF scheme for periodic data transmission in Smart Meter Network with cognitive radio

In order to solve the scalability issue for periodic data transportation, the authors in [10] propose a modified PCF scheme with the aid of cognitive radio technology (CR-PCF), in which the Smart Meters are allowed to use the white space to report the periodic data to the Local Collector as secondary users.

http://dx.doi.org/10.5772/62392

In the initialization phase of this algorithm, all the Smart Meters and Local Collector stay in the dedicated channel for the control signalling exchange. Local Collector senses all the candidate channels in the white space and then detects the available channels not occupied by Primary Users. After that, it distributes all the available channels to its associated Smart Meters with the broadcasting Poll message. After receiving the message, the Smart Meters having been allocated transmission resources are tuned to the allocated channels and transmit their packets to the Local Collector. Then, the Local Collector checks whether the received packets have been corrupted due to the collision with Primary Users. At the same time, it senses the current available channels and distributes them to the Smart Meters for the next round transmission together with the acknowledgement. It is noted that all the Smart Meters tune back to the dedicated channels so as to be able to receive the acknowledgement and Poll messages. This operation runs until all the Smart Meters finish the transmission of all the pending packets.

On the other hand, the fairness among all the Smart Meters is also quite important in the protocol design. For example, the Local Collector first serves the specific several Smart Meters in a generic group and does not allocate channels to the rest of the Smart Meters until the first several Smart Meters finish their transmission of all their packets. After that, the Local Collector focuses on serving other Smart Meters until they finish their pending packet transmissions. As a consequence, there may exist a situation that the number of Smart Meters which still need to be served is smaller than the number of available channels when approaching the end of contention-free frame. It is obvious that this case is a waste of channel resource, which will also lead to increasing the duration to finish the entire reporting transmission. In order to avoid this situation, we propose a Least Completed First Served Principle as shown in [10]. In brief, at each round of channel sensing and allocation, the Local Collector prefers to allocate the channels to the Smart Meters with the least completed transmissions so as to try to guarantee the fairness among all the Smart Meters.

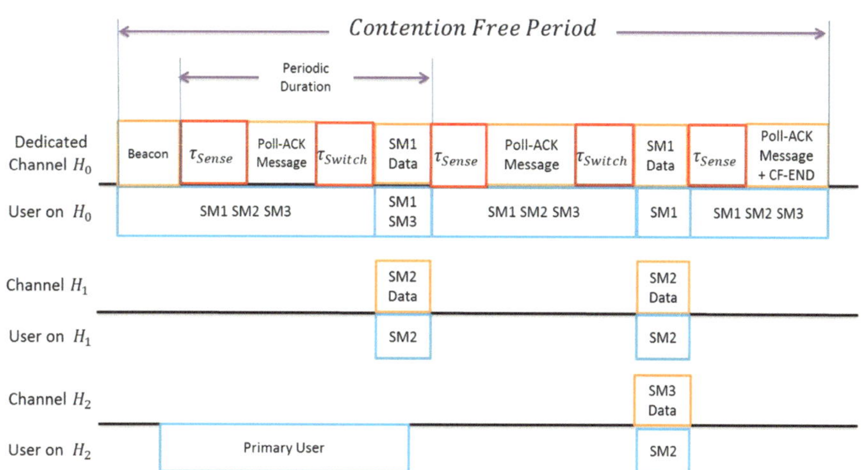

Figure 11. CR-PCF scheme operation example.

The detailed protocol scheme is exemplified in **Figure 11**. In addition, through comprehensive numerical and simulation in [10], the modified PCF with cognitive radio is shown to significantly outperform the traditional one in Smart Meter Network.

7. Conclusion

In this chapter, we briefly introduced the Smart Meter Network architecture and candidate communication technologies, which is quite important to the MAC protocols' design of Smart Meter Network. After that, we highlighted several significant MAC protocol design metrics and the associated research issues in the Smart Meters Network, including different data types, scalability, delay, fairness, expandability, security, etc. In order to solve the challenges and issues, we proposed two MAC protocol design solutions, grouping-based schemes and enhanced PCF scheme with the aid of cognitive radio. Furthermore, a short summary of MAC protocol improvement in IEEE 802.11ah TG is also included.

Author details

Yue Yang

Address all correspondence to: yueyang@uw.edu

University of Washington, Seattle, WA, USA

References

[1] Khalifa T., Naik K., and Nayak A., "A Survey of Communication Protocols for Automatic Meter Reading Applications," IEEE Communications Survey and Tutorials, vol. 13, no. 2, Second Qtr, 2011, pp. 168–182.

[2] Yang Y., and Roy S., "PMU Deployment for Three-Phase Optimal State Estimation Performance," IEEE International Conference on Smart Grid Communications, Vancouver, Canada, 2013.

[3] Yang Y., and Roy S., "PMU Deployment for Optimal State Estimation Performance," IEEE GLOBECOM 2012, Anaheim, CA, USA, 2012.

[4] Gungor V.C., Sahin D., Kocak T., and Ergut S., "A Survey on Smart Grid Potential Applications and Communications Requirements,"IEEE Transactions on Industrial Informatics, vol. 9, no. 1, 2011, pp. 28–42.

[5] Wang W., Xu Y., and Khanna M., "A Survey of the Communications Architectures in Smart Grid," Computer Networks, vol. 55, no. 15, 2011, pp. 3604–3629.

[6] Park B., Hyun D., and Cho S., "Implementation of AMR System Using Power Line Communication, IEEE/PES Transmission Distribution Conf. Exhibition, Yokohama, Japan, Oct. 2002.

[7] Peizhong Y., Iwayemi A., and Zhou C., "Deveploing Zigbee Deployment Guideline Under WiFi Interference for Smart Grid Applications," IEEE Transactions on Smart Grid, vol. 2, no. 1, 2011, pp. 110–120.

[8] Yarali A., "Wireless Mesh Networking Technology for Commercial and Industrial Customers," Proceeding Electrical Computing Engineering CCECE, Niagara Falls, Ontario, Canada, May 1–4, 2008, pp. 000047–000052.

[9] IEEE 802.11ah TG, "Specification Framework for TGah," IEEE 802.11-11/1137r15, May 2013.

[10] Yang Y., and Roy S., "PCF Scheme for Periodic Data Transmission in Smart Metering Network with Cognitive Radio," IEEE GLOBECOM 2015, San Diego, CA, USA.

[11] Yang Y., and Roy S., "Grouping Based MAC Protocols for EV Charging Data Transmission in Smart Metering Network," IEEE Journal of Selected Area on Communications, vol. 32, no. 7, 2014, pp. 1328–1343.

[12] Yang Y., Yin Y., and Hu Z., "MAC Protocols Design for Smart Metering Network," Automation, Control and Intelligent Systems, vol. 3, no. 5, 2015. pp. 87–94.

[13] Lewis R.P., Igic P., and Zhongfu Z., "Assessment of Communications Methods for Smart Electricity Metering in the UK," IEEE PES/IAS Conference of Sustainable Alternative Energy (SAE), Valencia, Spain, Sept. 28–30, 2009.

[14] Gungor V.C., Sahin D., Kocak T., Ergut S., Buccella C., Cecati C., and Hancke G.P., "Smart Grid Technologies: Communications Technologies and Standards," IEEE Transactions on Industrial Informatics, vol. 7, no. 4, 2011, pp. 529–539.

[15] Tan H., Lee H., and Mok V., "Automatic Power Meter Reading System Using GSM Network," IEEE International Conference on Control Application, Singapore, Oct. 1–3, 2007.

[16] AEIC Team, "AEIC/Smart Grid AMI Interoperability Standard Guidelines for ANSI C12.19 Communications," Association of Edison Illuminating Companies, Birmingham, AL, USA Version 2.0, November 19, 2010.

[17] IEEE 802.11ah TG, "Potential Compromise for 802.11ah Use Case Document," IEEE 802.11-11/0457r0, Mar. 2011.

[18] IEEE 802.11ah TG, "TGah Channel Model Proposed Text," IEEE 802.11-11/0968r3, Nov. 2011.

[19] Proakis J.G., "Digital Communications, Fifth Edition," McGraw Hill Publishing House, New York, NY, USA, 2009.

[20] Electric Power Research Institute, "Summary of EPRI Test Circuits," www.epri.com.

[21] IEEE 802.11ah TG, "Association ID Management for TGah," IEEE 802.11-11/0088r1, Jan. 2011.

[22] Ma H., Alazemi H., and Roy S., "A Stochastic Model for Optimizing Physical Carrier Sensing and Spatial Reuse in Wireless Ad Hoc Networks," IEEE International Conference on Mobile Adhoc and Sensor Systems, Washington DC, USA, Nov.7, 2005.

[23] IEEE 802.11 WG, IEEE 802.11e/D5, "Draft Supplement to Standard for Telecommunications and Information Exchange between Systems—LAN/MAN Specification Requirements—Part 11: Wireless Medium Access Control and Physical Layer Specifications: Medium Access Control Enhancements for Quality of Service," 2003.

[24] Kurose J., and Ross K., "Computer Networking A Top-Down Approach, Fifth Edition," Pearson Education Inc, New York City, NY, USA, 2010.

[25] Aust S., and Prasad R., "IEEE 802.11ah: Advantages in Standards and Further Challenges for Sub 1 GHz Wi-Fi," 2012 IEEE International Conference on Communications, Ottawa, Canada, Jun. 10–15, 2010.

Section 3

Platforms for Renewable Resources and Electronics in Smart Systems

Chapter 6

Power Electronics Platforms for Grid-Tied Smart Buildings

Mahmoud Amin

Additional information is available at the end of the chapter

http://dx.doi.org/10.5772/62920

Abstract

Renewable energy sources (such as sun, wind, water, or fuel cells) are attracting great interest for either grid-tied or off-grid arrangements in smart green buildings. It must be either used when generated, stored for future use on-site, delivered to the power grid, or shared among combination of these. Grid-tied buildings are connected to the utility grid service lines. Off-grid buildings have no connection to utility service lines. Both types employ inverters to convert power from direct current (DC) to alternating current (AC), and most off-grid systems have batteries to store energy for use when needed. Accordingly, power electronics systems are playing an important role as the enabling technology for smart grid. In addition, smart meter represents the interface part between the green building and the utility grid. In order to realize the interaction between both systems, a bidirectional power conditioning module is needed. This chapter introduces the different power electronics platforms suitable for grid-tied smart green buildings (such as residential homes, commercial, and industrial) as well as its integrative functionality with advanced metering infrastructure (AMI). In order to show the superiority of these platforms in conjunction with smart meters, a hardware case study with one of the most popular power electronics topologies is presented.

Keywords: power electronics, smart meters, green buildings, grid-tied systems, net meters

1. Introduction

In the future smart green buildings, there will be increasing connection to the distribution network of renewable energy sources, electric vehicles, and heat pumps. Most of alternative energy sources generate direct current (DC) power (for example, photovoltaic and fuel cell systems). As a result, a power electronics inverter is required to connect them into the utility grid. On the other hand, the alternating current (AC)-based alternative energy sources (for example, wind, wave, and hydro systems) can be either directly connected to the AC grid or indirectly through AC-DC and DC-AC conversion interface stages. Power electronics converters are then represents rapid independent control means of real-reactive power to satisfy grid-connected alternative energy conversion system needs [1]. In this chapter, we will explore the different power conversion topologies and energy link integration methodologies based on the renewable energy system structure (single or hybrid).

Grid-tied buildings are classified as either feed-in-tariff (FIT) or net metered (NM). In FIT systems, utilities purchase renewable energy at variable rates, which are usually higher than the sales price. This is tracked by using two meters: one to measure electricity going to the grid, and another for electricity coming from the grid. NM systems buy and sell electricity at the same rate, using a single meter, which runs either forward or reverse, depending on the direction of power flow. Most NM systems do not provide homeowners with credit for any electricity they generate beyond what they use. FIT systems, however, provide 100% credit for power put into the grid, allowing homeowners to receive a check from their utility when their production exceeds their use. In USA, regulations for grid-tied systems are established by the local municipality or state. Not all utilities allow FIT or net metering [2].

In fact, both FIT and NM systems can operate with either analog or digital (smart) meter. It became feasible to most people that smart meters are considered an integral part of any intelligent-based green building. In the past, analog regular meters can only provide distributors with power flow data at the substation level. With smart meters, it can deliver detailed in-depth real-time information about load energy consumption which extends visibility down to the consumer level. In addition, smart meter can help to manage and control customer loads remotely by involving more IQ functionalities into metering system design [3].

2. Home smart grid

In general, smart grid is considered an integrated and interactive power network. It brings generating units closer to consumers. On the other hand, consumers may also act as generators. A smart home involves three new components: smart control and measuring devices, digital communications systems, and computer software programs [4, 5]. **Figure 1** shows the smart grid simple structure that allows power to be fed into it from different energy sources and provides real-time management to maximize efficiency. The home smart grid makes individual homes energy efficient by reducing total energy consumption and lowering peak demand. Homeowners can install renewable energy generating components to supplement power they draw and feed power back when production exceeds home requirements. Home energy efficiency requires three devices: smart meter that measures power in real time, home receiver that allows homeowner to monitor power use over time, and power usage monitor (watt meter) that accurately measures power consumption by individual devices.

http://dx.doi.org/10.5772/62920

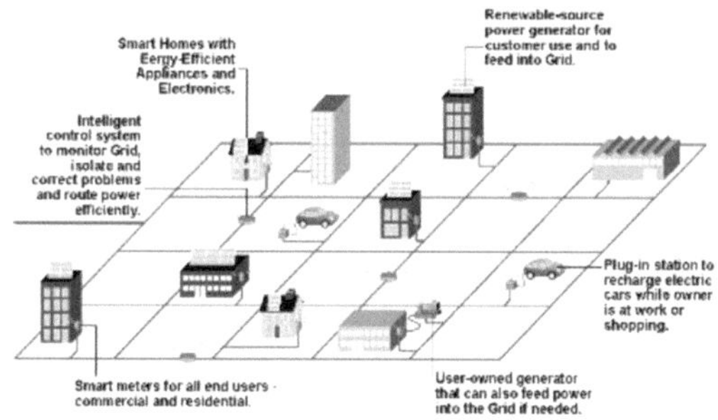

Figure 1. Home smart grid infrastructure.

2.1. Home meters

There are two common types of metering home systems: analog and digital meters, as shown in **Figure 2**. The analog (electric, gas, and water) meter measures flow over a given period of time but cannot transmit measurement data. Utility companies are unable to charge according to production costs. Users have no incentive to minimize use during peak demand. Digital metering system represents an advanced smart meter that can be used to monitor the energy consumption remotely and transfer this information to the control center through secured communication network [5]. In order to facilitate monitoring functionality, we can add a digital display to show the energy consumption and its corresponding cost to the customers. The in-home display (IHD) is mainly used for this purpose. Another type of smart meters is the power strip one. It contains one or more channels with voltage/current transducers to monitor the amount of the power usage and relay contact to control major equipment in the building [6–8]. The smart meter has the following features:

- Read electric use at least once per hour

- Transmit data to electric company

- Monitor for disturbances or interruptions

- RF- or IP-based wireless network

- Net meters allow homeowners to produce electricity and sell excess to utility company

Figure 2. A typical residential analog and digital (smart) electric meter with a digital display showing cumulative kilowatt hours used.

2.2. Smart meter and smart building

Giving everyone a smart meter will not deliver a smart building. Although smart meters are an essential component of a truly smart home, there is more to smart buildings than just smart meters (hardware, software, communication links, and controllers). It is called advanced metering infrastructure (AMI) that measures, analyzes, and regulates energy usage. The IHD plays an important role in real-time information monitoring. It may communicate directly with appliances. It also provides simple tools for cutting energy consumption by 5–10% [9, 10]. **Figure 3** shows different popular IHD units that allow electrical consumers to read data from their home's smart meter and other measuring devices.

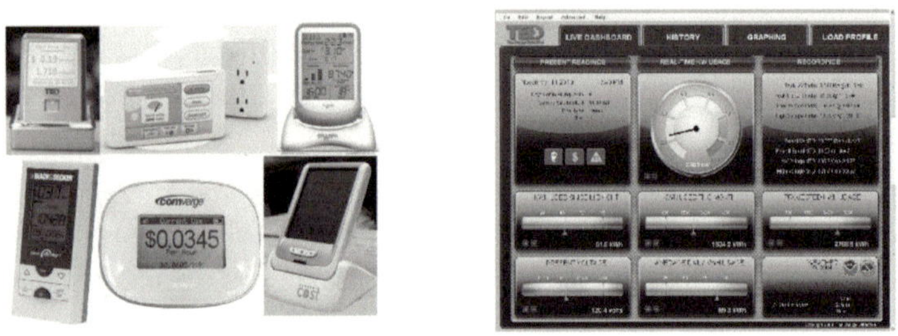

Figure 3. The PowerPortal Home display unit and real-time power, cumulative use, and other data.

3. Renewable energy sources

Renewable energy sources are being developed in many countries to reduce CO_2 emissions and provide sustainable electrical power. The balance of particular technologies and their scale changes from country to country. However, hydro, wind, biomass, tidal stream, and photovoltaic (PV) are common choices. Substantial equipment cost investment is generally recovered and cost savings are realized over time [11, 12]. **Figure 4** shows photographs for different renewable energy sources.

Figure 4. Renewable energy sources.

3.1. Power converters platforms

Power electronics represents the enabling technology for green alternative energy sources in smart buildings. It involves sophisticated conversion topologies as an interface integral part between smart power distribution network and local microgrid sources [5]. The installation of the new advanced alternative energy technologies (e.g., PV and wind) is playing a vital role in residential and commercial smart buildings. Typically, installed capacity ranges from few kilowatts for distribution-customer level to several megawatts for high-voltage transmission grid. The following characteristics are important for power electronics systems for smart grids: high efficiency, optimal energy transfer, bidirectional power flow, high reliability, synchronization capabilities, EMI filtering, smart metering, real-time information, communications, and fault tolerance/self-healing. **Figure 5** shows different power electronics layers to integrate a cluster of prosumers (an entity in the future grid capable of both producing and consuming electric power) into the smart building.

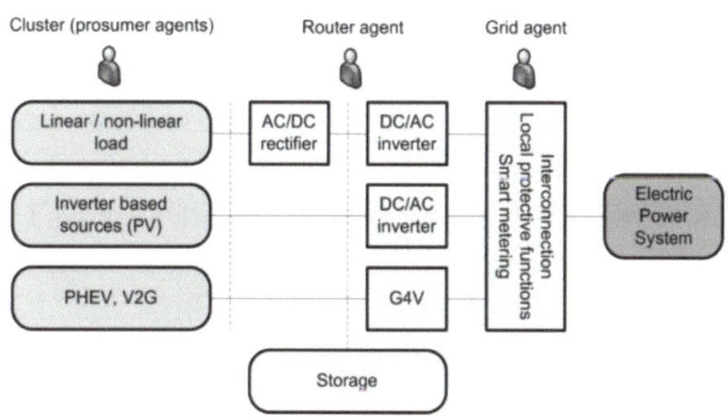

Figure 5. Power electronics control and communication structure in smart buildings.

Power electronics converters can operate renewable sources on either off-grid standalone basis or grid-connected basis. Grid-tied platforms can be classified into two main categories: single input single output (SISO) platform and multi-input single output (MISO) platform. As an example, a building that involves only one DC renewable energy source such as PV modules mounted on the roof, then, the platform structure needed to connect the grid-tied FIT building is SISO as shown in **Figure 6**. The output of a PV system is DC and therefore a DC-AC converter is essential for grid connection.

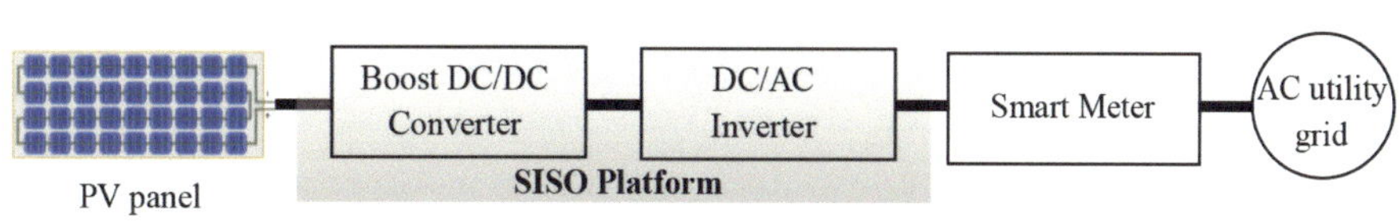

Figure 6. DC source-based SISO platform for grid-tied FIT system.

Another SISO platform can be used if the source was AC-based variable speed turbines such as wind, small hydro, and tidal power generation as shown in **Figure 7**. Typically, this platform uses two stages of power conversion. The first stage is AC-DC to convert from the variable frequency AC generator power output to controlled/uncontrolled DC. The second stage is then DC-AC to convert from DC into synchronized 50/60 Hz AC utility grid. To capture the maximum power, the turbine rotational speed is set to the optimum range.

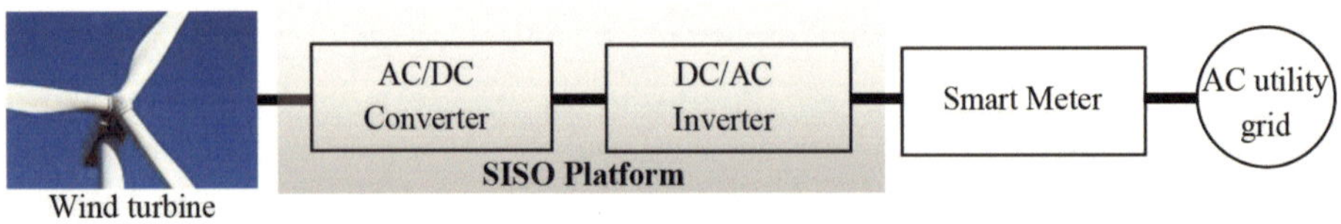

Figure 7. AC source-based SISO platform for grid-tied FIT system.

There are several other renewable technologies that use steam or gas turbines with synchronous generator such as biomass energy. This technology can be directly connected to utility grid and is not discussed in this chapter. In general, power converter interface units can be used to control the amount of the reactive power injected from renewable energy systems into main AC grid. Then, network voltage is controlled while managing active power to satisfy the requirements of the utility grid. On the other hand, **Figure 8** shows hybrid connectivity to provide larger energy production to the building as well as the utility grid which is called as MISO platform. Additionally, MISO platform is sometimes called as multi-port power converter (MPPC) since it combines more than one input port based on the number of the micro energy sources and storage installed in the building.

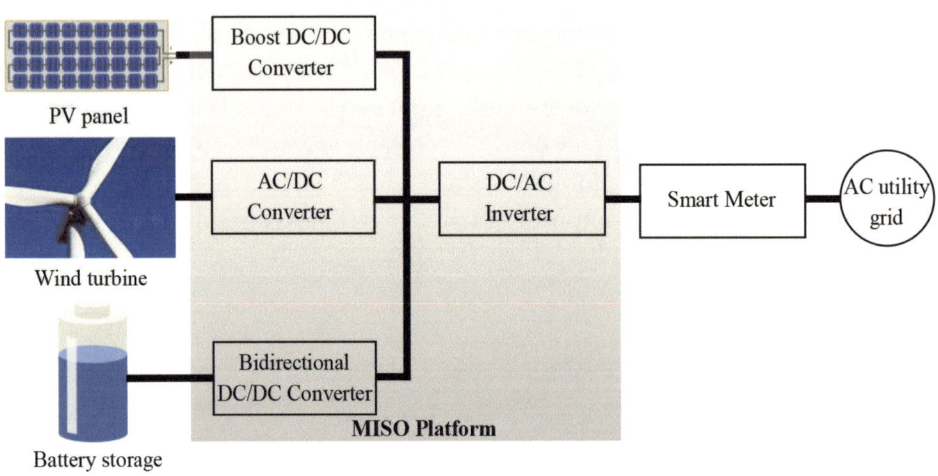

Figure 8. Hybrid source-based MISO platform for grid-tied FIT system.

http://dx.doi.org/10.5772/62920

3.2. PV system in home

Energy from sun is transformed into an energy source that can do useful work or provide useful heat. We pay cost of converting stored sun energy into useable energy form. Most energy sources originated from sun except geothermal energy. The sun's free energy can be utilized in three forms:

- Passive use; for example, allowing natural sunlight in for heat.
- Direct heat transfer; for example, water heater collects solar heat and transfers it to water
- PV effect; causes some substances to generate an electric current when exposed to sunlight

Many countries such as USA, Spain, and Germany, have extended their PV system installation to larger capacities. FITs are finding great interest with PV as it provides guaranteed payment per energy unit (p/kWh) generated from local renewable power sources. Professional help is essential to install a grid-connected solar power system in home. It requires "balance-of-system" equipment such as power conditioning, safety, meters, and instrumentation. On the other hand, installing a standalone PV power system is usually chosen for homes in remote areas. Successful systems employ a combination of power-generating technologies and techniques to reduce electric energy requirements. Standalone system is still not as complex as creating grid-connected system. **Figure 9** shows both grid-connected and standalone PV systems in home. The grid-connected generating system feeds power into the grid through one meter while the home's electric current is brought into the home from the grid through another meter. The standalone solar generating system with PV panels connected through a controller to a battery stack and an inverter.

3.2.1. Maximum power point tracking (MPPT) control

Figure 10 represents the maximum output power locus for an individual PV cell. It is interesting to observe that neither the maximum voltage nor the maximum current are the same for different levels of illumination at the maximum power point. As the output power (Po) is given as:

$$P_o = V_o I_o \tag{1}$$

Then the peak power may be given from the following condition:

$$\frac{dP_o}{dV_o} = V_o + I_o \frac{dV_o}{dI_o} = 0 \tag{2}$$

that is:

$$\frac{dV_o}{dI_o} = -\frac{V_o}{I_o} \tag{3}$$

The meaning of this expression is that the dynamical internal resistance of the source should match the external load resistance, leading to special power peak tracking control approaches.

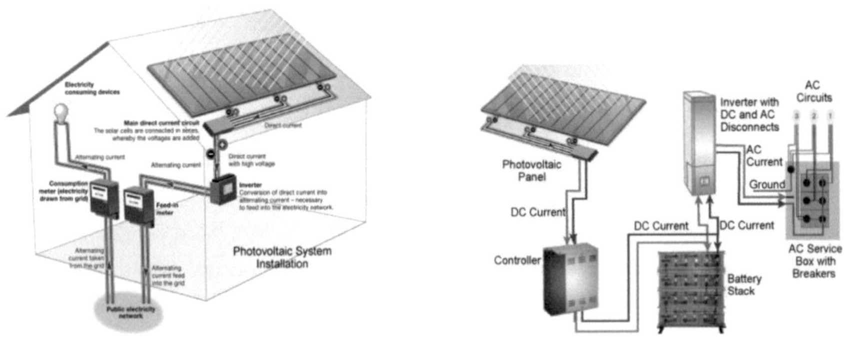

Figure 9. A schematic diagram shows installation of a grid-connected and standalone PV power system in home.

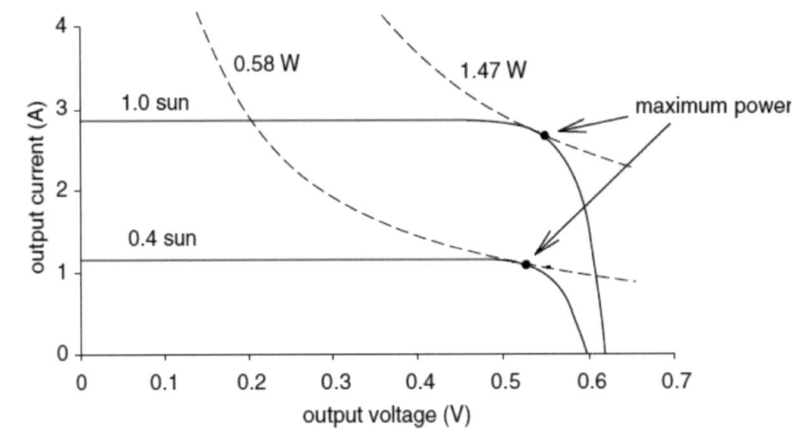

Figure 10. Conditions of maximum power for various illumination (insolation) levels.

3.2.2. PV-based SISO platform

Figure 11 shows the schematic of SISO-based grid-connected PV system. The main elements of this system are:

1. Boost DC-DC converter to increase the output voltage and extract the max. power
2. Single-phase DC-AC voltage source inverter (VSI)

3. Grid interface output filter with/without isolation transformer

4. Voltage/power controller

There are several other DC-DC conversion topologies that can be used in the first stage such as flyback, half bridge, full bridge, and push-pull [13]. Generally, the DC-DC converter output is maintained to be constant as an input of the inversion stage. On the other hand, the MPPT technique is continuously used to find the proper PV voltage that allows most power to be extracted while PV system parameters (for example, insolation and cell temperatures) changes. The DC voltage obtained from the DC-DC converter is inverted to 50/60 Hz AC. A VSI is widely used. Typically, VSI uses a pulse width modulation (PWM) switching pattern to reduce the output harmonic contents. Then, a final stage of grid-tie-filter is connected between VSI output and AC grid in order to minimize the current harmonics injected into main power network. An isolation transformer is sometimes placed at the VSI output to prevent DC injection into the grid.

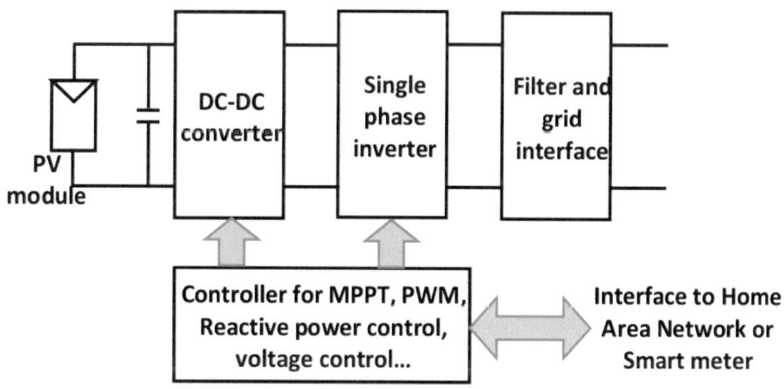

Figure 11. Domestic grid-connected SISO PV system block diagram.

3.3. Wind energy system in home

Wind is a manifestation of solar radiation. Air currents are winds that power weather systems continuously across the surface of the planet. The annual average wind speed is larger than or equal to 12 mph considering terrain and wind gusts. Favorable locations have wind resource potential of 510 kWh or more. We should also consider other factors such as investment, zoning codes, location, and so on. For home-scale wind turbines, residential typical range is 500 W–100 kW. It is determined based on amount of electricity to generate, budget, and wind resource. The 5–15-kW turbine should provide 75–100% of the typical home's power requirements. **Figure 12** shows how a wind turbine works for small-scale 2-kW home model and the main components for grid-connected wind turbine generating system [14].

According to Betz theory, the turbine mechanical power (P_t) is given by:

$$P_t = 0.5 C_p \rho A V^3 \quad kg \cdot m / s \tag{4}$$

where ρ is the air density, C_p is the power coefficient, V is the wind speed, and A is the area of the rotor blades. Considering the wind speed, wind generation system (WGS) can be classified into no, partial, and full load conditions. The wind turbine operates in the no load condition if the measured wind speed is not within the cut-in/cut-out wind speed range. On the other hand, WGS output power is regulated at full load power region by changing the pitch angle. In addition, the maximum extracted power can be achieved by controlling the tip speed ratio. As the rotor speed must change according to the wind intensity, the speed control of the turbine has to command low speed at low winds and high speed at high winds, so as to follow the maximum power operating point as indicated in **Figure 13**. It is observed that the maximum power output occurs at different generator speeds for different wind velocities [15].

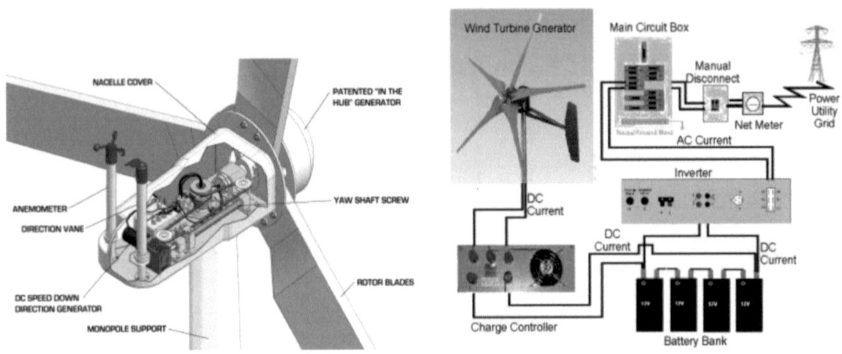

Figure 12. Hummer Wind Power's 2 kilowatt wind turbine and components for a grid-connected wind turbine generating system.

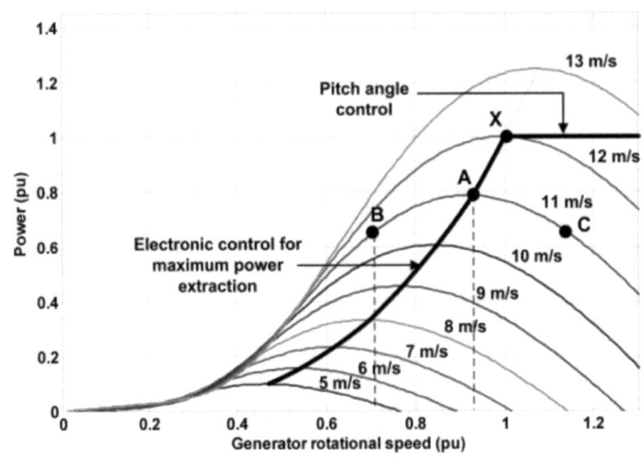

Figure 13. Turbine rotation versus power characteristic related to wind speed.

http://dx.doi.org/10.5772/62920

3.3.1. Wind turbine generators

For variable-speed wind turbine (VSWT), there are several types of generators that can be used. **Figure 14** shows the doubly fed induction generator (DFIG) which employs a wound rotor induction machine with back-to-back converter-inverter connected to its rotor terminals. The real-reactive power flow of the rotor circuit is controlled via power converters. The rotor side converter is also used to change the rotor speed through the active power absorbed or injected into rotor. Furthermore, the wind generator may be self-excited induction (SEIG) or external excited synchronous (EESG) machine with power converter. The maximum power is captured by controlling the generator speed while stator output frequency varies with wind speed conditions. The first stage of the power converter is used to convert the variable frequency power to DC power. Then, the DC power is inverted to 50/60 Hz AC grid power via the second conversion stage. Additionally, multi-pole permanent magnet synchronous generator (PMSG) represents another wind turbine design with direct-driven gearless advantage. It introduces better mechanical efficiency as well as higher power density.

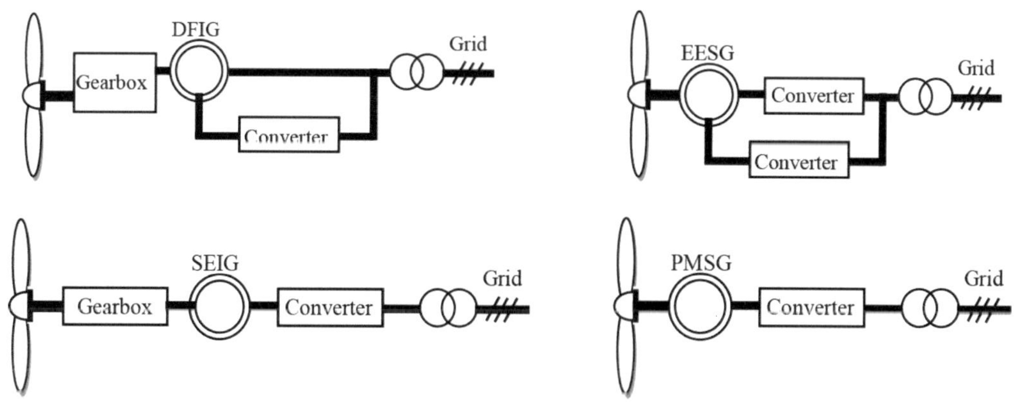

Figure 14. Variable-speed wind turbine generation technologies.

3.3.2. Wind-based SISO platforms with SEIG

Figure 15 shows a simple structure of the SEIG-based SISO platform for wind generation system. The system contains three cascaded stages (AC-DC, DC-DC, and DC-AC). The first stage employs a diode bridge rectifier to convert the variable generator AC output into unregulated DC. Then, the variable DC is converted to constant voltage DC link as an input of the inverter stage. However, diode rectifier stage causes undesirable effects of generator operation through voltage/current low-order harmonic distortion. Another option is to use two stages back-to-back converter/inverter as shown **Figure 16**. The first stage uses a PWM controlled rectifier which achieves high power factor and less harmonic distortion in the generator side. However, this structure needs the use of bootstrap driver circuits as the occurrence of short circuit through each phase is possible. **Figure 17** shows another variation of modified back-to-back power electronic converter structure that only uses four switches in each stage instead of six switches. On the other hand, the use of this topology is limited for loss optimization due to voltage balance problem across DC-link capacitor and less power capability.

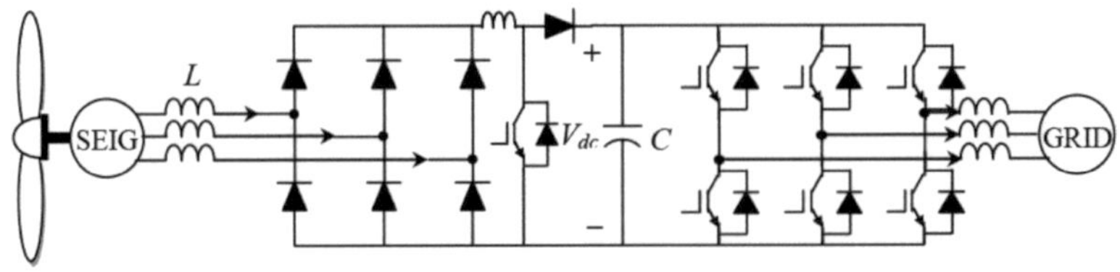

Figure 15. WECS with power factor correction using DC-DC stage.

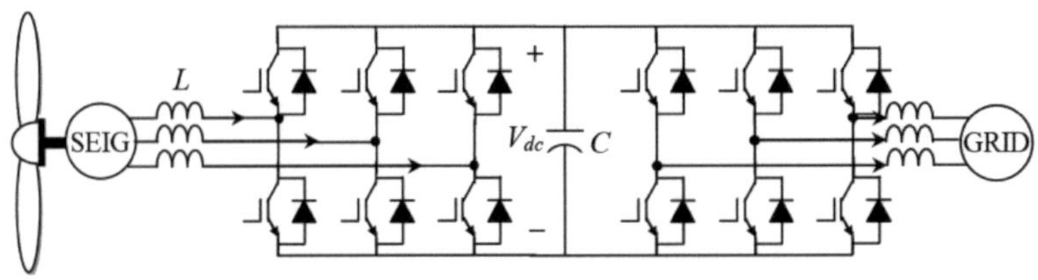

Figure 16. WECS with back-to-back converter.

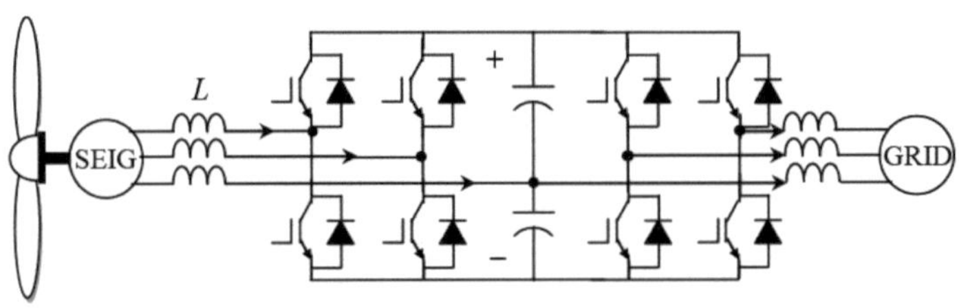

Figure 17. WECS with modified back-to-back converter.

3.3.3. Wind-based SISO platforms with PMSG

Low-speed high-torque PMSGs are used as the preferred solution in variable-speed high-power direct-driven wind generators. However, low-speed operation produces low voltage profile at the generator terminals. As the electrical power available from the wind generation system cannot be delivered directly to the grid, power electronics plays a decisive role in overcoming this limitation. **Figure 18** shows a bridge boost rectifier (BBR) in cascaded with VSI topology. A step-up transformer is used for grid connection; since VSI produces low voltage profile due to low-speed direct-driven generator operation. This topology has same features as the topology that was previously shown in **Figure 15** in addition to the grid-tie-transformer.

http://dx.doi.org/10.5772/62920

Another option to achieve high power factor in the generator side is to use a voltage source converter (VSC) in cascaded with VSI topology in the WECS as shown in **Figure 19**. The transformer stage still required to boost the inverter output voltage level to the grid voltage level. In **Figure 20**, a semi-controlled rectifier topology was connected to achieve same VSC topology advantages with higher efficiency; since less switching losses can be obtained through reducing the number of switching devices. However, it will produce larger current harmonic distortion because only one half cycle is controlled. The common drawback for all previous configurations is the use of the transformer for grid connection which increases the size of the power converter circuit. Then, this structure becomes feasible for small WECSs.

On the other hand, the same rectifier topologies can be efficiently utilized with current source inverter (CSI) topology feasible for low-speed direct-driven PMSGs. **Figure 21** shows same configuration as the other in **Figure 18** but with CSI working as a boost inverter stage. This configuration has the same drawbacks as illustrated previously. However, omitting the step-up transformer represents an additional advantage which reduces the size and weight of the system.

In **Figure 22**, the VSC topology can be used with CSI topology in order to achieve same advantages as the configuration in **Figure 19**. Dual boost feature is obtained with low harmonic distortion. However, larger losses are expected by increasing the number of switching devices (12-switches) which reduce the overall efficiency. The CSI increases the voltage towards the mains by itself, so the output voltage of the VSC must be lower than the lowest rectified line-to-line voltage.

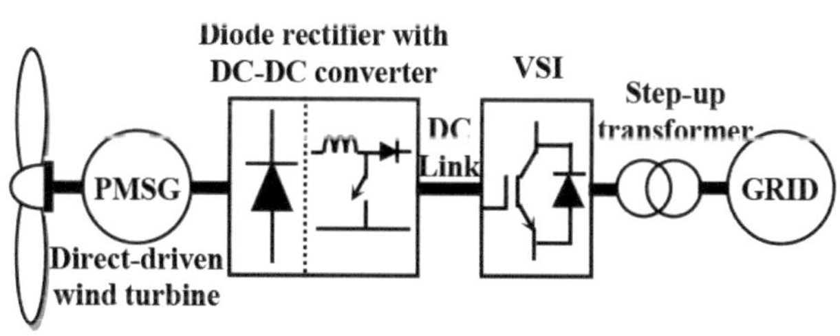

Figure 18. PMSG with BBR and VSI structure.

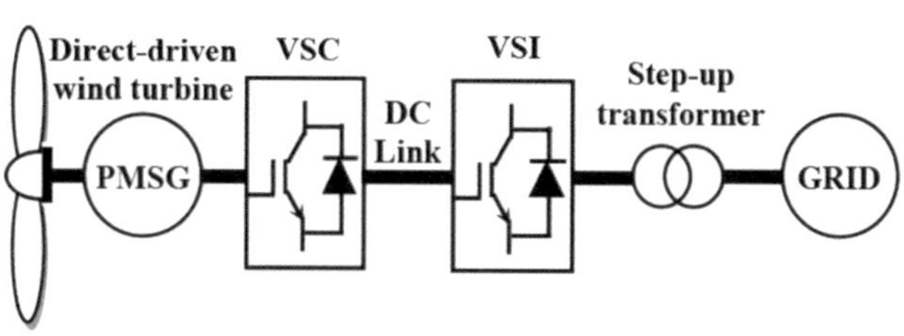

Figure 19. WECS with VSC-VSI converter structure.

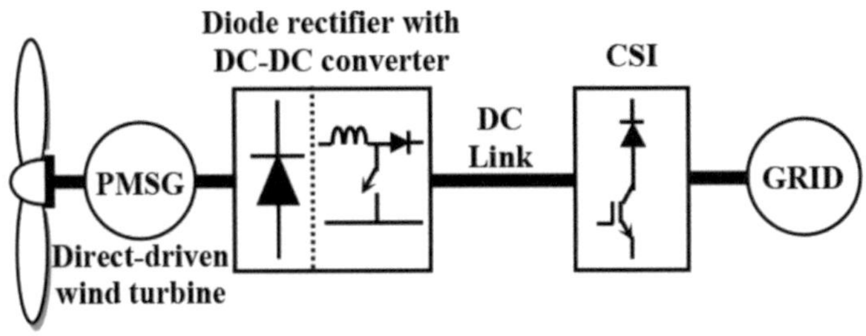

Figure 20. WECS with semi-controlled rectifier and VSI structure.

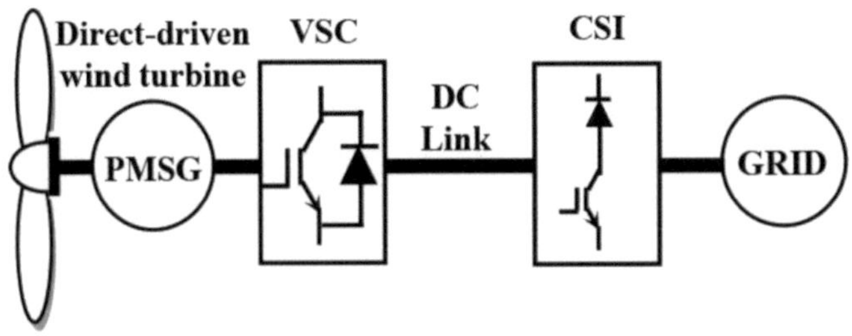

Figure 21. WECS with BBR stage and CSI structure.

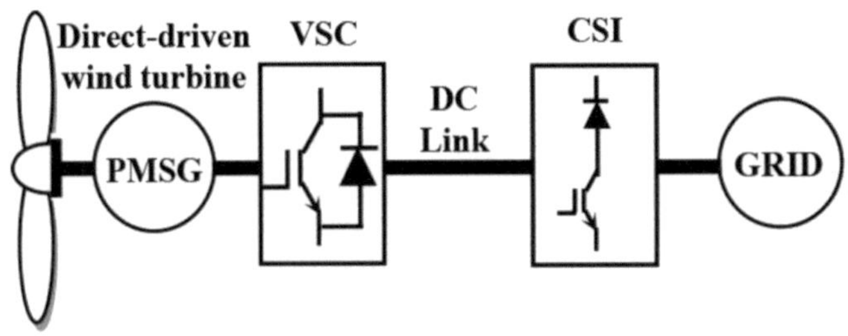

Figure 22. WECS with VSC-CSI structure.

3.4. Hybrid energy system in home

In hybrid energy systems (wind, PV, fuel cell, and battery storage) and its application such as plug-in hybrid vehicles (PHEVs), MISO converters became widely used to enhance the stability and reliability. **Figure 23** shows an application example for the MISO (sometimes called multi-input boost converter MIBC) platform feasibility with PHEVs. In MISO, the AC grid is connected to the DC-Bus via boost rectifier, the PV/fuel cell is connected to the DC-Bus via boost converter and the energy storage system is connected to the DC-Bus through a bidirectional DC/DC converter.

http://dx.doi.org/10.5772/62920

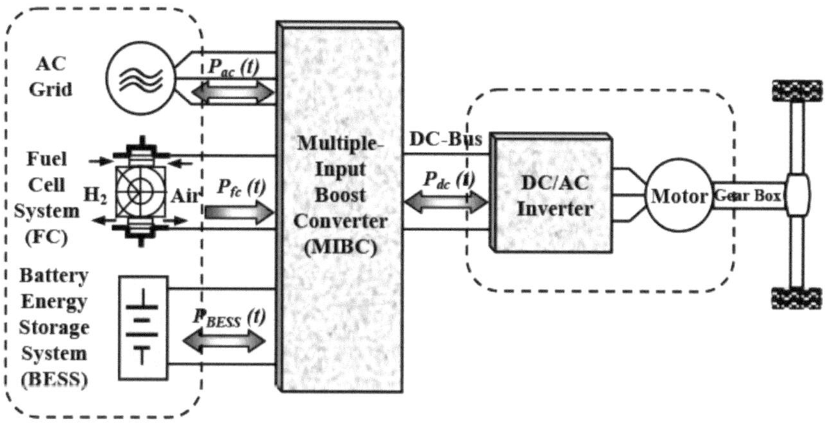

Figure 23. The block diagram of the PHEV drive train with MISO structure.

3.4.1. MISO with vehicle to grid (V2G)

PHEVs have an advantage of AC power grid connectivity compared to self-contained hybrid electric vehicles (HEV). The vehicle-to-grid (V2G) can be utilized when the vehicle battery is in the discharge mode while grid-to-vehicle (G2V) is used when the battery is in the charge mode. The V2G mode is considered as a promising concept for electric power grid stability via large storage vehicular system. However, the availability of electricity supplies to recharge vehicle battery is still a significant challenge to enable this concept. **Figure 24** shows an example of the V2G system components. The system involves six main parts: (1) AC power grid and alternative energy resources; (2) master-based independent control; (3) charging stations; (4) bidirectional power flow and point-to-point Plug-in electric vehicle (PEV) communication; (5) smart metering and control system; and (6) the vehicular technology including its battery charging management.

In general, PEVs with V2G interfaces can charge or inject energy into the grid when parked and connected. The concept requires three elements: a power connection to the grid, a communication connection with the grid operator, and suitable metering. Communications must be bidirectional to report battery status and receive commands. In fact, the current challenge is how to make smart metering system aware of the battery's state of charge (SOC) and its capacity in real time. In [16], smart metering system with both on/off boards was proposed to support V2G concept. As a result, the PHEV can represent a controllable load while integrating it with green energy sources via smart meter information. GPS locators are also useful in the large-scale real-time energy management operation. In addition, field area network is used to realize monitoring and communication functionalities among PHEV charging stations.

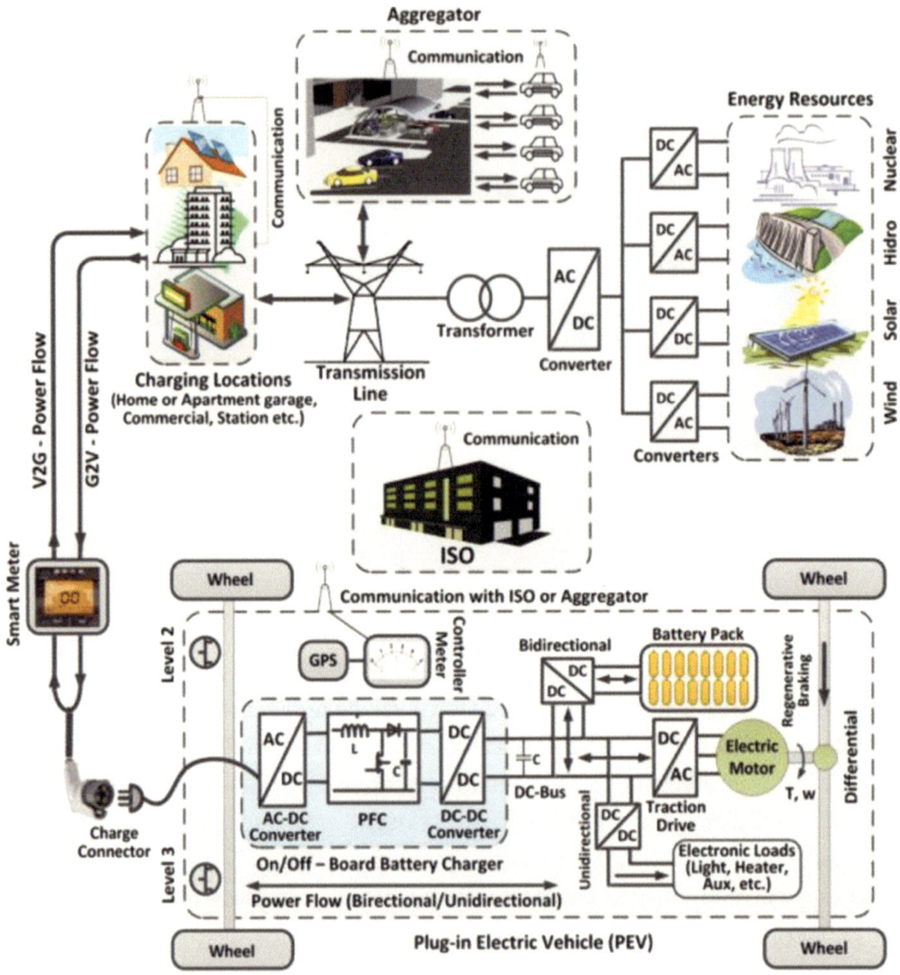

Figure 24. V2G system with smart meter.

3.4.2. SISO with wireless charging electric vehicle (EV)

Another feasible option for EV is that we can use the simple SISO platform to operate a wireless power transfer charging station as shown in **Figure 25**. There are several stages for wireless charging of EV. Firstly, an AC-DC rectifier with PFC ability is used to convert the AC utility grid power into DC power source. Secondly, the converted DC power is then inverted into a high-frequency AC power as the transmitting primary coil. The insulation failure of the primary coil is possible. As a result, a high-frequency isolation transformer may be connected between the inverter output and the primary side of the transmitting coil to ensure charging system protection. The magnetic field of the primary transmitting coil induces an AC voltage on the secondary receiving EV coil. Finally, the secondary transmitted AC power is then rectified via AC-DC converter to charge the EV battery [17].

http://dx.doi.org/10.5772/62920

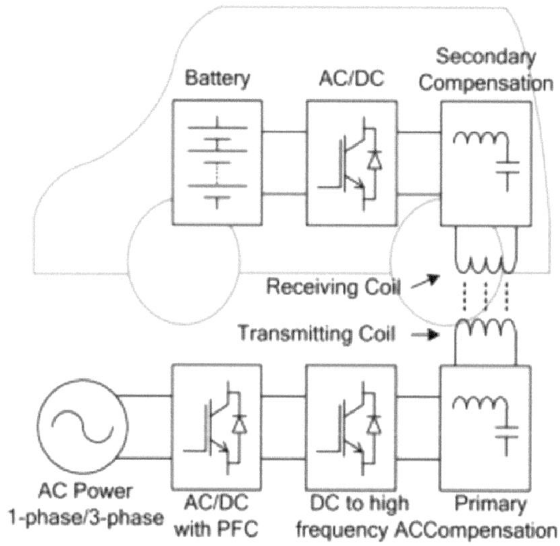

Figure 25. Typical wireless power transfer EV charging system.

4. Case study

VSI is considered one of the most efficient power conversion topologies for AC/DC-source-based MISO platform. A bidirectional power flow control strategy is suggested here to enable the smart operation between local building and electric utility service lines. The primary intent of the inverter development with smart functionalities is to enable an efficient interconnection and economical operation for dispersed PV-based building installations to the utility grid. Some distinctive aspects of this case are smart metering, the provision of pricing information to consumers, the provision of some control options to consumers, and information exchange on a fully networked system enabled by massively deployed sensors [18]. It is in this regard that the inverter with the aforementioned smart functionalities is being proposed in this section.

The PV panels provide the input of the DC-DC converter with 132 V. Then, the step-up converter boosts the PV voltage level to 350 V_{dc} which serves as the input of the smart DC-AC inverter. Moreover, the inverter setup is a part of MISO system structure that involves lead-acid battery energy storage bank with 160 V nominal voltage and 100 Ah rated capacity. The battery storage is connected to the DC link via a bidirectional buck-boost DC-DC power converter.

The case study was performed through Matlab/Simulink software program and real-time experimental hardware implementation in order to verify the validity of this platform with smart meters for grid-tied FIT systems. The test results were acquired through a laboratory-scale MISO frame hardware setup in the sustainable energy systems (SES) laboratory at Manhattan College. The overall system schematic and photograph is shown in **Figure 26**. The system comprises multiple converters used to control power flow under harsh ambient conditions. Different operating modes are possible, for example, the system can be supplied with electric power by way of a 3-phase wind generator, a PV, and batteries. The system is housed in an air-cooled case (with thermal protection) and communicates with the master controller (TMS320F2812 DSP platform) via a D-connector bus. The signal interface features analog and digital I/Os to allow for the connection of a wide variety of sensors, for example, temperature sensors, resolver inputs.

A voltage/current sensor module is connected through the AC-bus common coupling point between the MISO system and utility grid. This module is used to acquire the power measurements as a part of smart meter functionality for real-time energy management operation. The stability of DC-bus voltage is the main parameter to decide on energy flow scenario. Then, we can set upper and lower limits for DC-bus voltage in order to maintain system balance operation while connected to AC grid. For example, if V_{dc} is higher than V_{dc_up}, then it means that there is excess power available from wind and PV generators. Accordingly, the battery will operate in the charging mode to store energy while supplying local load demand and sending power back to grid based on smart building contract agreement. In this case, net meter measurement will show a negative reading which indicates feeding energy into utility grid. On the other hand, if V_{dc} is lower than V_{dc_low}, then it means that either one or both wind and PV generators is supplying less power into MISO system. As a result, the battery will operate in the discharging mode to support local load demand while receiving a portion of the needed power from grid. In this case, net meter measurement will show a positive reading which indicates receiving energy from utility grid. Finally, if V_{dc} is within the upper and lower stability limits, then it means that both wind and PV generators are balanced and sharing power with utility grid to supply building loads. As a result, the battery will operate in the rest mode which improves battery life time. The detailed parameters and specifications of MISO platform are listed in **Tables 1** and **2**.

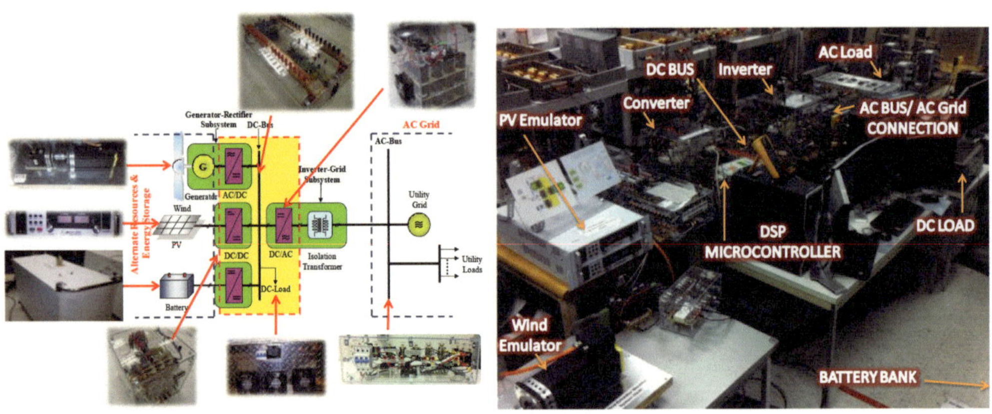

Figure 26. The schematic diagram and photo of the MISO experimental hardware test setup.

http://dx.doi.org/10.5772/62920

4.1 Pure *P-Q* injection capability

An inverter grid-tie system has been designed to verify the effectiveness of the MISO injection capability for AC power into the grid. The inverter has been tested under various power factor and command step conditions with 12 mH grid filter. The switching frequency of the devices is set to be 5 kHz. The grid voltage is 120 V, 60 Hz. The system controller and PWM generation are conducted by DS1103 PPC Controller Board. **Figure 27** shows the grid-tie inverter control under pure active power, pure leading, and lagging reactive power conditions. The output is 5 kW active power (0°), 2.5 kVar lag reactive power (−90°), or 5 kVar lead reactive power (90°).

4.2 Hybrid *P-Q* injection capability

Figure 28 shows the grid-tie inverter control with equal active power and reactive power command. The active power is 3 kW with either lagging (1.5 kVar) or leading reactive power (3 kVar). We can notice that the grid voltage-current angles are always larger than 0° and smaller than 90° (lag or lead). The testing results and measurements are confirming the validity and superiority of the developed MISO platform with smart inverters interconnecting green buildings to utility grid network.

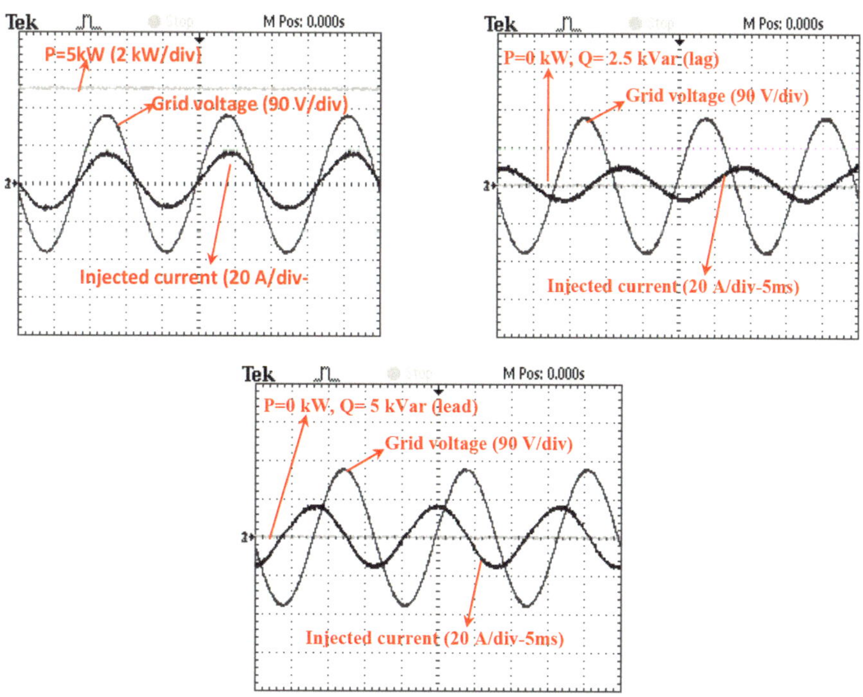

Figure 27. Pure active-reactive power injection test results for 5 kW active, 2.5 kVar lag reactive, and 5 kVar lead reactive powers.

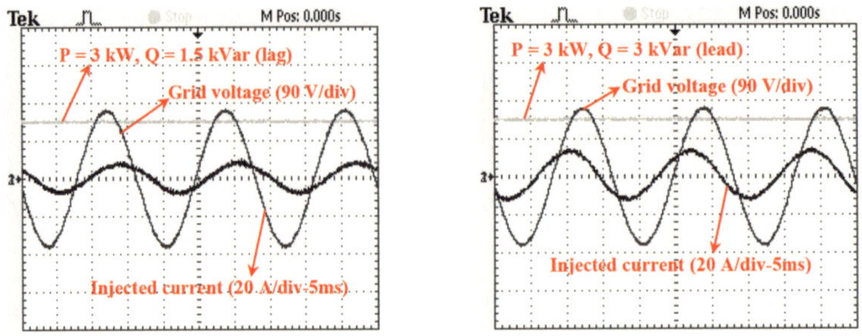

Figure 28. The test results for the hybrid active-reactive power injection for 3 kW active with 1.5 kVar lag reactive powers and 3 kW active with 3 kVar lead reactive powers.

5. Conclusion

In this chapter, a detailed study of the different power electronics platforms for grid-tied smart buildings was presented. Both analog and digital (smart) metering systems were discussed including FIT and NM infrastructures. The main power electronics converter platforms were categorized into two types as SISO and MISO structures. Each platform was introduced in regards to several types of alternative energy resources (wind, PV, etc) involving battery storage and its use in our home applications. PHEV technology with (V2G-G2V) concepts and its wireless power transfer operation were discussed. A hardware experimental case study with MISO platform involving smart bidirectional grid-tied inverter was performed to verify the validity of the developed power electronic structure for smart systems. The new power electronics platforms were studied to prove its superiority for grid-tied smart metering systems.

Symbol	Quantity	Value
f_s	Grid frequency	60 Hz
V_s	Grid voltage	120/208 Vrms
V_{dc}	DC-link voltage	350 V
C_{dc}	DC-link capacitance	1200 μf
L_f	Filter inductance	12 mH
f_{sw}	Switching frequency	5 kHz
kp^i	Proportional current controller	10
ki^i	Integral current controller	100
kp^c	Proportional voltage controller	0.07
ki^c	Integral voltage controller	0.7

Table 1. PV and battery storage specifications and control parameters.

PV		Battery bank	
Parameter	Specification	Parameter	Specification
Output power	0–6 kW	Rated voltage	160 V
OC and SC points	(10 A, 140 V)	Connection	10-series
Max, operating point	(9 A, 132 V)	Rated capacity	100 AH
Switching frequency	50 kHz	Battery type	Lead-acid
Input inductor	0.33 mH	Total power	12 kW

Table 2. Smart bidirectional inverter specifications and control parameters.

Author details

Mahmoud Amin*

Address all correspondence to: mahmoud.amin@manhattan.edu

Department of Electrical and Computer Engineering, Manhattan College, Riverdale, NY, USA

References

[1] Ekanayake J, Liyanage K, Wu J, Yokoyama A, and Jenkins N. Smart Grid, Technology and Applications. 1st ed. Wiley; 2012. 306 p, West Sussex, United Kingdom. DOI: 10.1002/9781119968696

[2] Kruger A and Seville C. Green Building: Principles and Practices in Residential Construction. 1st ed. Delmar Cengage learning; 2012. 544 p, Clifton Park, NY, USA. DOI: 10.1002/ep.11805

[3] Beard C. Smart Grids for Dummies. Wiley; 2010. 68 p, West Sussex, United Kingdom. ISBN: 978047066537–4

[4] Wells Q. Smart Grid Home (Go Green with Renewable Energy Resources). 1st ed. Delmar Cengage Learning; 2012. 304 p, Clifton Park, NY, USA. ISBN: 9781111318512

[5] Simoes M, Roche R, Kyriakides E, Suryanarayanan S, Blunier B, McBee K, Nguyen P, Ribeiro P, Miraoui A. A comparison of smart grid technologies and progresses in Europe and the U.S. IEEE Transactions on Industry Applications, vol.48, no.4, pp.1154–1162, July–August 2012. DOI: 10.1109/TIA.2012.2199730

[6] Hyun S, Yamazaki T, Minsoo H. Determining location of appliances from multi-hop tree structures of power strip type smart meters. IEEE Transactions on Consumer

Electronics, vol.55, no.4, pp.2314–2322, November 2009. DOI: 10.1109/TCE.2009.5373804

[7] Benzi F, Anglani N, Bassi E, Frosini L. Electricity smart meters interfacing the households. IEEE Transactions on in Industrial Electronics, vol.58, no.10, pp.4487–4494, October 2011. DOI: 10.1109/TIE.2011.2107713

[8] Dinesh C, Nettasinghe B, Godaliyadda R, Ekanayake M, Ekanayake J, Wijayakulasooriya J. Residential appliance identification based on spectral information of low frequency smart meter measurements. IEEE Transactions on Smart Grid, vol.PP, no.99, pp.1–12, October 2015 DOI: 10.1109/TSG.2015.2484258

[9] Seunghyun P, Hanjoo K, Hichan M, Jun H, Sungroh Y. Concurrent simulation platform for energy-aware smart metering systems. IEEE Transactions on Consumer Electronics, vol.56, no.3, pp.1918–1926, August 2010. DOI: 10.1109/TCE.2010.5606347

[10] Nian L, Jinshan C, Lin Z, Jianhua Z, Yanling H. A key management scheme for secure communications of advanced metering infrastructure in smart grid. IEEE Transactions on Industrial Electronics, vol.60, no.10, pp.4746–4756, October 2013. DOI: 10.1109/TIE.2012.2216237

[11] Amin M and Mohammed O. Modern Technologies in Energy Efficiency Improvement: Design, Control, Implementation, and Applications. Scholars' Press; 2013. 308 p, Deutschland, Germany. ISBN: 9783639705379

[12] Farret F and Simoes M. Integration of Alternative Sources of Energy. Wiley; 2006. 504 p, Hoboken, New Jersey. DOI: 10.1002/0471755621

[13] Li W and He X. Review of nonisolated high-step-up dc/dc converters in photovoltaic grid-connected applications. IEEE Transactions on Industrial Electronics, vol.58, no.4, pp. 1239–1250, April 2011. DOI: 10.1109/TIE.2010.2049715

[14] Amin M and Mohammed O. Development of high-performance grid-connected WECS for optimum utilization of variable speed wind turbines. IEEE Transactions on Sustainable Energy, vol.2, no.3, pp. 235–245, July 2011. DOI: 10.1109/TSTE.2011.2150251

[15] Amin M and Mohammed O. DC-bus voltage control technique for parallel-integrated permanent magnet wind generation systems. IEEE Transactions on Energy Conversion, vol.26, no.4, pp. 1140–1150, Dec. 2011. DOI: 10.1109/TEC.2011.2163409

[16] Yilmaz M and Krein P. Review of the impact of vehicle-to-grid technologies on distribution systems and utility interfaces. IEEE Transactions on Power Electronics, vol. 28, no.12, pp.5673–5689, Dec. 2013. DOI: 10.1109/TPEL.2012.2227500

[17] Siqi L and Mi C. Wireless power transfer for electric vehicle applications. IEEE Journal of in Emerging and Selected Topics in Power Electronics, vol.3, no.1, pp.4–17, March 2015. DOI: 10.1109/JESTPE.2014.2319453

[18] Souza W, Marafao F, Liberado E, Diniz I, and Serni P. Power quality, smart meters and additional information from different power terms. IEEE Transactions in Latin

America, (Revista IEEE America Latina), vol.13, no.1, pp.158–165, Jan. 2015. DOI: 10.1109/TLA.2015.7040643

Chapter 7

Estimating the Photovoltaic Hosting Capacity of a Low Voltage Feeder Using Smart Meters' Measurements

Vasiliki Klonari, Jean-François Toubeau,
Jacques Lobry and François Vallée

Additional information is available at the end of the chapter

http://dx.doi.org/10.5772/62418

Abstract

Maximizing the share of renewable resources in the electric energy supply is a major challenge in the design of the future energy system. Regarding the low voltage (LV) level, the main focus is on the integration of distributed photovoltaic (PV) generation. Nowadays, the lack of monitoring and visibility, combined with the uncoordinated integration of distributed generation, often leads system operators to an impasse. As a matter of fact, the numerous dispersed PV units cause distinct power quality and cost-efficiency problems that restrain the further integration of PV units. The PV hosting capacity is a tool for addressing such power system performance and profitability issues so that the different stakeholders can discuss on a common ground. Photovoltaic hosting capacity of a feeder is the maximum amount of PV generation that can be connected to it without resulting in unacceptable power quality. This chapter demonstrates the usefulness of smart metering (SM) data in determining the maximum PV hosting capacity of an LV distribution feeder. Basically, the chapter introduces a probabilistic tool that estimates PV hosting capacity by using customer-specific energy flow data, recorded by SM devices. The probabilistic evaluation and the use of historical SM data yield a reliable estimation that considers the volatile character of distributed generation and loads as well as technical constraints of the network (voltage magnitude, phase unbalance, congestion risk). As a case study, an existing LV feeder in Belgium is analysed. The feeder is located in an area with high PV penetration and large deployment of SM devices.

Keywords: distributed generation, low voltage network, probabilistic simulation, smart meters, photovoltaic

1. Introduction

Maximizing the share of renewable resources in the electric energy supply is a major challenge in the design of the future energy system. Concerning the low voltage (LV) distribution system, this objective aims at increasing the self-sufficiency of LV feeders, based on local resources, while responding to the climate change. In such feeders, distributed photovoltaic (PV) generation is the mostly met distributed energy resource (DER).

So far, the biggest share of distributed PV units came with no previous planning or reinforcement of the network while monitoring data in the residential or commercial sector were absent almost everywhere in Europe. Given the lack of controllability in common LV networks, the uncoordinated integration of PV units often leads to distinct power quality issues. As a result, the connection of new PV units and therefore the increase of renewable energy share slows down. Adding to this fact the growing volatility of electricity consumption in the distribution network, the adoption of a streamlined planning approach for the future energy system becomes urgent.

In this evolving framework, distribution system operators (DSOs) are called to safeguard a stable and secure power supply in all possible demand conditions while fostering the massive integration of DER generation. In cost-efficiency terms, this fact highlights the necessity of leaving behind deterministic worst case planning approach. This traditionally applied approach focuses on the least favourable network operation states, which are very rare. Naturally, it leads to very restrictive decisions in terms of PV hosting capacity or to costly network reinforcements.

Given the current uncertainty of DSO costs and revenues, new planning tools are required for considering the constant variability of the energy network [1]. This argument becomes even more solid in view of the upcoming integration of electric vehicles and the development of flexibility services. As a matter of fact, both are seen as basic components of the future energy model. The large deployment of smart metering (SM) devices in the residential and commercial sector will drastically enlarge the potential of cost-effective planning approach. Indeed, customer-specific data will result in a better insight of the power distribution system.

Considering the above facts and the probabilistic character of the EN 50160 technical standard [2], [3] (which addresses the LV network), this chapter presents a feeder- and customer-specific probabilistic method that estimates the DER hosting capacity of an LV feeder. Practically it introduces a probabilistic tool that uses customer-specific energy flow data recorded by SM devices (that are installed in the studied feeder), over a period of 2 to 4 years depending on the customer. The deployment and probabilistic elaboration of long-term energy measurements (SM readings) yield a reliable estimation that considers the volatile character of distributed generation and loads as well as several operational metrics.

Section 2 of this chapter outlines some of the existing scientific contributions that address this subject and presents the drivers for developing the proposed methodology and algorithm.

Section 3 presents the overall structure of the developed algorithm and Section 4 thoroughly describes the primordial contribution of customer-specific SM readings in this development. Section 5 explains the computation process of the maximum acceptable PV hosting capacity.

As a case study, a real LV feeder in Belgium is analysed. The feeder is located in an area with high PV penetration and large deployment of SM devices. When the probabilistic character of EN 50160 standard's voltage limits is considered, the estimated PV hosting capacity is proved to be much higher than the one obtained with a deterministic approach, based on worst case energy flow profiles. At the same, the use of the SM readings verifies the computation of the technical constraints in the feeder that cannot be considered with a probabilistic approach (violation of the maximum current capacity of the lines).

2. Current Framework

In many regions worldwide, DER integration is hampered due to slow or over rigid hosting capacity review processes. As a result, customers who want to invest and play an active role in managing their energy usage are increasingly unable, in expediency and cost-efficiency terms, to do so. In this context, a stream-lined approach together with the expansion of allowable DER integration approvals seem to be a necessity [4].

However, the expansion of allowable approvals depends heavily on DER admissible penetration levels, which are determined by local DSOs. In order to increase penetration levels while facilitating the application review process, DSOs should incorporate automated DER hosting capacity analyses. An example process flow for incorporating such analysis into the DER integration review process is outlined in **Figure 1**.

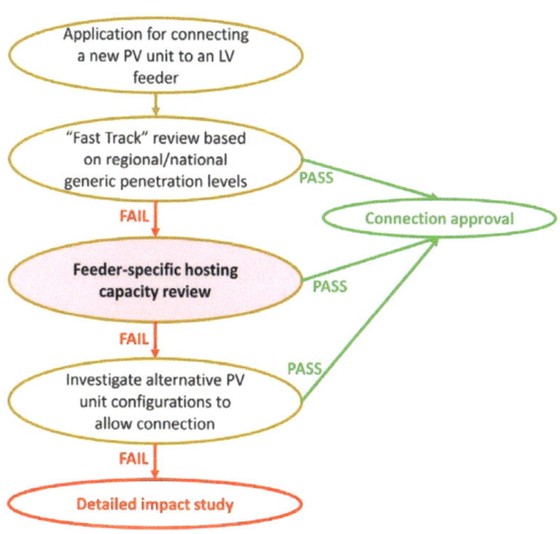

Figure 1. Process flow for incorporating hosting capacity analysis into the DER integration process.

Currently, many energy utilities are adapting their DER hosting capacity review so as to remove or update restrictive maximum allowable limits [5]. To this end, the Electric Power Research Institute (EPRI) presents a set of models that could be used by DSOs or electric utilities [6], [7]. These feeder-based methodologies are very solid computation examples that take account of all steady state operational criteria.

Focusing on PV hosting capacity, the EPRI's report presents stochastic analysis as a highly appropriate tool for determining feeder hosting capacity for distributed PV units. The stochastic deployment concerns the position and size of future PV units while the steady state estimation of the feeder is done with deterministic approach. Indeed, the analysed state estimation scenarios are based on four worst case load/ PV generation profiles.

In the same vein, a set of studies addressing the European framework and the EN 50160 standard highlight the efficiency of stochastic and probabilistic analysis in determining hosting capacity or otherwise the impact of PV generation in LV feeders [5]–[12].

Meanwhile, the European Photovoltaic Industry Association (EPIA) and the technical standard EN 50160 suggest that distribution networks should be designed on a probabilistic basis. For example, EN 50160 standard deals with the voltage characteristics of LV feeders in probabilistic terms. It gives recommendations that, for a percentage of measurements (e.g. 95%) over a given time, the voltage value must be within specified limits.

Most of the existing methodologies deploy the stochastic analysis regarding the size and position of PV units and not the load/generation profiles of customers. However, the ongoing integration of SM devices in LV networks enlarges the potential of using feeder-specific or even customer-specific data for modelling energy flows. According to [9], performing long-term measurements in the LV network is highly valuable and strongly recommended, not only for estimating the maximum PV hosting capacity, but also for voltage coordination of the network in general. Measurements of the voltage magnitude for a large time of customers may be time-consuming and expensive; however, many countries have already been installing energy meters that also allow the recording of voltage, current, active power and other metrics.

Considering these facts, the EPRI's report [7] estimates PV hosting capacity using feeder-specific data to create either absolute worst case scenarios (maximum recorded generation-minimum recorded load) or load/PV time-of-day coincident worst case scenarios. Therefore, feeder-specific data are indeed used; however, the steady state estimation of the feeder is still done with a deterministic approach. Consequently, this approach does not consider the fact that the time-of-day in which worst case values apply for a specific customer does not necessarily coincide with the one of other customers connected to the same feeder. Nevertheless, the operational criteria of the feeder are determined both by the individual user's demand and by the simultaneous demands of other network users. Since the demands of every user and the degree of coincidence between them constantly vary, so does the operation of the feeder [3].

The above argument demonstrates that although customer-specific SM data are primordial for creating reliable network models, there is another challenge that needs to be addressed. The latter lies in the fact that customers follow volume-wise (kWh) or capacity-wise (kW) an almost stable daily pattern. However, this pattern does not necessarily remain the same on the time axis. In long-term decision making, profiles should be based on the recorded ones considering all possible deviations. Those deviations could be inserted either as random statistical errors or by making random possible combinations of the recorded values or even by combining both approaches.

Consequently, reliable models that use customer-specific real SM readings and take into account load/PV time- and customer-variability are necessary for applying a less conservative and more cost-effective hosting capacity review. Probabilistic and particularly Monte Carlo approach are very suitable to address this modelling challenge.

3. The PV Hosting Capacity Computation Tool

Hosting capacity is defined as the maximum amount of PV that can be accommodated in the feeder without impacting system operation (reliability, power quality, etc.) under existing control and infrastructure configurations [7]. This chapter presents a tool that uses probabilistic state estimation, 15-min customer-specific SM energy flow readings and feeder-specific technical parameters to estimate the PV hosting capacity of a given LV feeder.

The proposed methodology aims to address the central block of **Figure 1** (*"Feeder-specific hosting capacity review"*) by providing a detailed location- and customer-specific DER hosting capacity analysis. The analysis takes into account the EN 50160 standard operational criteria [2], [3]. In particular, the focus is on voltage magnitude and unbalance which are the primary technical concerns in LV feeders with distributed PV generation. The maximum line capacity is also taken into account so as to address important reverse power flows due to high PV injection.

Although the EN 50160 standard sets the same voltage limits in all European countries (except from cases where stricter limits are locally imposed), the maximum line capacity heavily depends on the respective DSOs. In certain countries, line sections are chosen based on a long-term strategy that aims at minimising voltage and congestion risk even if loads and generation increase importantly in the future. However, such approach leads to higher initial investment which is not necessarily cost-effective. In other cases, line sections are chosen based on actual conditions or short term future scenarios so that customised solutions are applied as soon as problems arise.

Apart from steady state constraint management, there are other considerations that could be accounted for, such as transformer aging factor, line losses, etc. Such criteria are usually considered in an overall cost-benefit analysis (CBA); however, at present they are not addressed by the EN 50160 standard. Depending on the country and the applied DSO tariff methodology ("cost-plus", "revenue cap", etc.), DSOs are incentivised to reduce certain operation costs that can or cannot be integrated in their tariffs. Thus, the impact of such criteria on decision making, varies in function of the distribution utility. For this reason, this chapter computes PV hosting capacity focusing on commonly adopted EN 50160 standard criteria and line capacity issues. Line losses in the feeder, during PV injection hours, are also addressed however their rise is not imposed as a constraint to the further increase of admissible hosting capacity.

3.1. Overview of the simulation tool

As previously said, this chapter presents a probabilistic algorithm that determines the PV hosting capacity of an LV feeder by elaborating feeder-specific SM measurements. The SM measurements are the necessary input for performing a reliable steady state analysis of various possible energy flow scenarios in the studied feeder. The flowchart in **Figure 2** presents the structure of the simulation algorithm, which is entirely developed in MATLAB®. The energy exchange scenarios are generated by the Monte Carlo algorithm sampling from the historic SM data of the feeder [13], [14]. The power flow analysis is performed with the three-phase algorithm that is presented in [14] and outlined in Appendix A. Both balanced and unbalanced situations can be considered in this study.

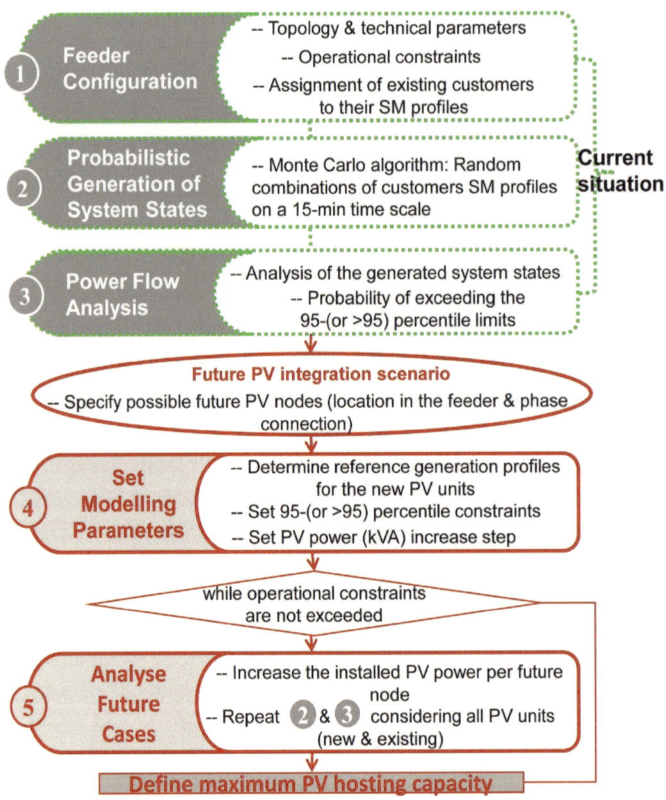

Figure 2. Flowchart of the PV hosting capacity computation tool.

3.2. Feeder model

The feeder model is constructed based on the technical parameters of the lines, the position of the customers, the installed PV power per node, the voltage at the MV/LV transformer secondary output and the respective set points and bandwidths in case voltage control algorithms are integrated. The feeder model also assigns the load/PV generation SM datasets to the respective customers. This necessary information is available to the DSO.

Regarding the PV hosting capacity computation, the possible future locations of the PV units have to be specified in the feeder model. This analysis is not based on stochastic random distribution of PV units along the feeder. A set of scenarios regarding the positions of future PV nodes is specified and each one of them is studied separately so as to focus on its specific impact on the feeder, which is possible thanks to the customer-specific SM datasets.

The technical constraints that must be respected for the current situation and for future scenarios are the ones specified in local, regional or national directives. However, these operational constraints can be determined in a more restrictive manner, depending on the case. In the EU framework, the steady state constraints are set by the EN 50160 standard. Regarding voltage magnitude and unbalance, 95-percentile limits are suggested. For considering this EU standard, the simulation tool verifies that the following criteria apply for the whole system (in current and future installed PV power scenarios):

$$P_{overvoltage}\left(V_{i,j} > 1.10 \cdot V_{nom}\right) < 0.05$$

$$P_{undervoltage}\left(V_{i,j} < 0.90 \cdot V_{nom}\right) < 0.05 \qquad (1)$$

$$P_{unbalance}\left(VUF_i > 2\%\right) < 0.05$$

where $P_{overvoltage}$, $P_{undervoltage}$ and $P_{unbalance}$ represent respectively the probability of having an overvoltage, an undervoltage or exceeding the phase voltage unbalance limit at node i (phase j) over a number M of simulated network states. In $V_{i,j}$, i stands for nodes 1 to N (total number of nodes in the feeder) and j stands for phase a, b or c. VUF_i stands for the Voltage Unbalance Factor at node i.

The thermal limits of the cables are also considered in the computation. The current carrying capacities of the lines should not exceed the DSO requirements or the recommended values in technical standards such as [15]. The load flow analysis of each system state is performed with the three-phase algorithm that is explained in the Appendix [14].

4. The use of SM measurements

4.1. Customer profiles and feeder state modelling based on historic SM datasets

The load/ PV profiles of existing customers are created by using their respective SM recorded datasets. Practically, each dataset consists of values of 15-min PV energy injection to the grid ($E_{inj,\ grid}$), 15-min PV energy generation ($E_{inj,pv}$) and 15-min energy consumption ($E_{cons,\ grid}$), recorded at the respective customer. For the 15-min time step q, the net energy consumption of customer i is computed by introducing the respective SM recorded values in the following formula:

$$E_{load\,i,q} = E_{cons,grid\,i,q} + \left(E_{inj,pv\,i,q} - E_{inj,grid\,i,q}\right) \quad (2)$$

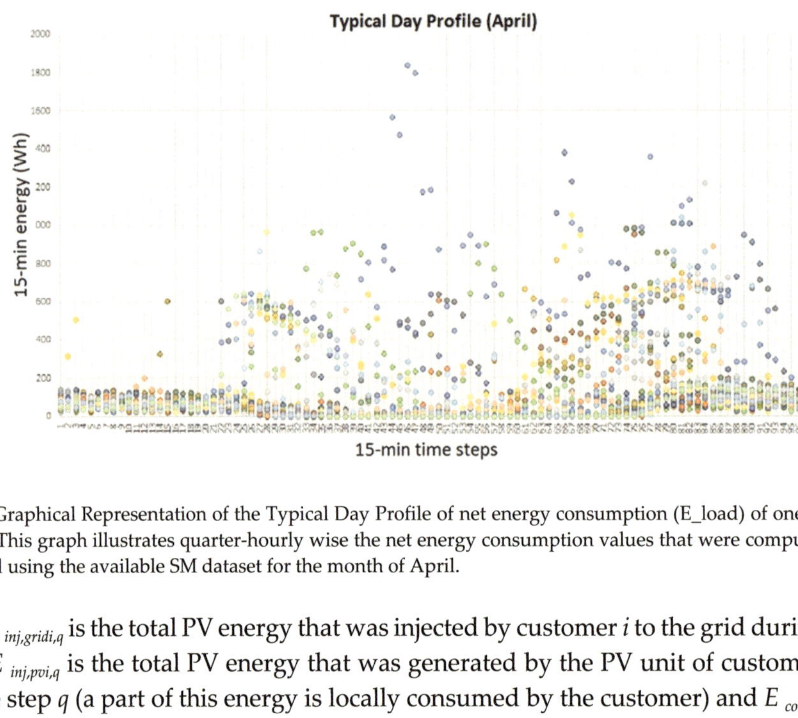

Figure 3. Graphical Representation of the Typical Day Profile of net energy consumption (E_load) of one PV customer for April. This graph illustrates quarter-hourly wise the net energy consumption values that were computed by applying (2) and using the available SM dataset for the month of April.

where $E_{inj,grid\,i,q}$ is the total PV energy that was injected by customer i to the grid during the time step q, $E_{inj,pv\,i,q}$ is the total PV energy that was generated by the PV unit of customer i during the time step q (a part of this energy is locally consumed by the customer) and $E_{cons,grid\,i,q}$ is the total energy consumption that was absorbed by customer i during the time step q. These three values are recorded by the SM device that is installed at customer i.

The two 15-min resolution datasets of $E_{inj,pv}$ and E_{load} (energy exchanged at the coupling point of the customer with the feeder) are used to build two Typical Day Profiles (TDPs) for each customer; the "typical day" represents and characterizes a selected period (which can be a month, a season, a year or so on). Each TDP reflects the variation that the respective parameter can have at every individual quarter of an hour of a "typical day".

The TDP of energy consumption E_{load} created for the month of April for one of the customers is graphically represented in Figure 3. Practically, if an SM device has been monitoring the energy flow of a user over a period of one month, each dot represents the total amount of net energy that was consumed by the user during time step q (q=1:96, represented on the horizontal axis) in one of the days of the month. Therefore, for each time step q, D dots (D= number of days of the month) have been drawn on the diagram of Figure 3, each one representing the energy consumption of the user during the respective time step in each one of the D days. Similar graphs can be created for all feeder users that are equipped with an SM device, both for their net energy consumption E_{load} and their PV energy generation $E_{inj,pv}$.

As a next step, the customer-specific TDPs are statistically transformed to 2 x 96 Cumulative Distribution Functions (CDFs) of Probability, one for PV generation and one for energy consumption (=2) multiplied by the number of quarters-hourly time steps in a day (=96), representing each one of the studied months. This transformation is made by applying the basic statistical formula, for each 15-min dataset:

$$CDF(X) = P(X \leq x) \tag{3}$$

Therefore, the "typical day" of each customer (for the respective month) can be illustrated by two diagrams like the one presented in **Figure 4**, one for PV injection and one for net energy consumption. Exactly the same methodology is applied to build TDPs for the r.m.s. voltage at the secondary output of the MV/LV transformer, using 15-min data, also recorded at the specific MV/LV substation.

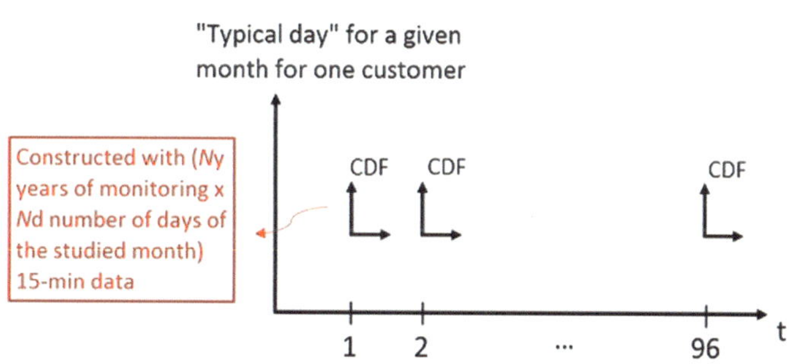

Figure 4. CDFs of probability are created for PV injection of each customer to the grid, net energy consumption of each customer and voltage at the MV/LV transformer output for each 15-min time step.

The created statistical distributions (CDFs) of the time varying parameters (PV injection, net energy consumption and voltage at the MV/LV transformer) are used for defining multiple possible network states corresponding to each 15-min time step. Practically, a MC algorithm is applied for randomly sampling the values of the variable parameters at each node of the studied LV feeder as explained in [13], [14] . The combination of the nodal sampled values defines each network state that will be afterwards analysed by the power flow algorithm

(Appendix). Subsequently, for each state, a feeder with N_{total} nodes is characterized by the following values:

$$\{E_{loadi}, E_{inj,pvi}, t_i\} \text{ for nodes } i = 2 : N_{total} \tag{4}$$

$$\{V_{MV/LV}\} \text{ for node } i = 1$$

where E_{loadi} is the 15-min net energy consumption at node *i*, $E_{inj,pv,i}$ is the 15-min PV energy generation at node *i*, t_i is the time repartition factor of the consumed or generated energy at node *i* and $V_{MV/LV}$ is the voltage at the MV/LV transformer node. In case there are no sufficient data for intra 15-min intermittencies of energy flow, it can be assumed that the power flow is stable during the 15 minutes of each time step. This means that the time repartition factor t_i corresponds to 15 minutes and is thus equal to 0.25.

The power flow analysis of the feeder requires considering each system state as instantaneous, and therefore the sampled energy values have to be transformed into instantaneous power values ($E_{inj,pv} \rightarrow P_{inj,pv}$ and $E_{load} \rightarrow P_{load}$). This means that for each node we can consider either power injection or power consumption since both of them cannot be applied simultaneously at an instant. In such a way, the instantaneous power value that represents the power flow at the point of common coupling (PCC) of each customer *i* with the feeder is determined as follows:

$$P_{i=} \frac{E_{loadi} - E_{inj,pvi}}{0.25} \tag{5}$$

If P_i is positive the respective customer *i* is instantaneously consuming power from the grid whereas if P_i is negative, the customer is instantaneously injecting power into the grid.

The probabilistic deployment of this simulation tool relies on the principle that load/PV generation profiles of customers are highly time-varying. The generation of the system states is therefore based on a very large number of random combinations of customers' energy flow values. This time-variability induces another variability that concerns the time coincidence of the load profiles of various customers. Both arguments are very important when assessing the impact of PV generation on a LV network. Indeed, the consideration of this variability, both in the time axis and regarding customers coincidence, makes more realistic the simulation of the network operation. Such an approach can lead to less restrictive and more cost-effective decisions that do not rely on rare extreme cases but on the most frequent ones.

4.2. Generation profiles of future PV nodes

A key component in accurately assessing the impact of future PV units is reliably representing their generation profiles. Based on the findings of several studies, geographically close customers are entirely correlated as far as their PV generation profiles are concerned [16]. For this reason, this study considers that the generation profiles of future PV customers will be very similar, along the time axis, to the ones of the existing PV units.

As previously explained, the load/PV generation profiles of customers with SM devices are made of 96 Cumulative Distribution Functions (CDFs) of probability built with the 15-min recorded datasets. Concerning PV generation, such CDFs are apparently not available for the future PV units. For this reason, the available SM datasets are used in this case to create a reference CDF, based on the 15-min generation SM datasets of the existing PV owners [17]. This reference CDF is used to simulate the time-variability of PV generation at the future PV nodes.

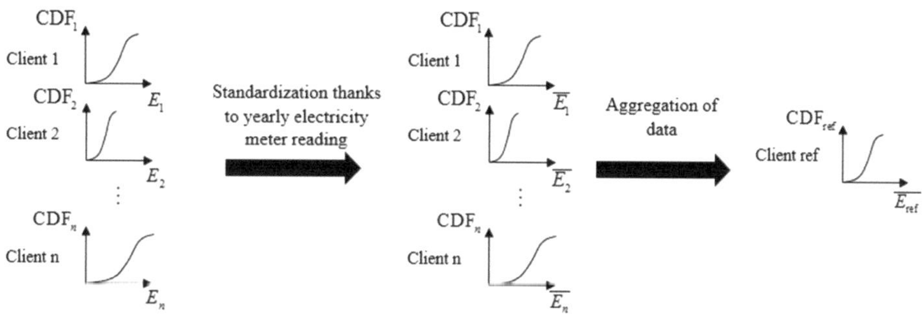

Figure 5. Construction of the reference CDF (CDF of the reference PV node) in an LV feeder with SM devices.

In reality, customers that are connected to the same LV feeder can have different PV units' sizes. Assuming an equivalent statistical distribution of their PV power profiles due to geographical proximity, the principle is to create a standardized reference CDF for PV generation in the specific feeder, based on the measurements of the available SM devices. Initially, the CDF for the 15-min PV energy generation $E_{\text{inj,pv,j,q}}$ of each existing PV node j is normalized by applying the following relation, for each time step q:

$$\overline{E_{inj,pv,j,q}} = \frac{E_{inj,pv,j,q}}{E_{tot,j}},, \text{for } j = 1 : N_{\text{SM}} \tag{6}$$

where N_{SM} is the number of users in the feeder that are equipped with an SM device, $\overline{E_{inj,pv,j,q}}$ values are the normalized 15-min energy generation values of customer j during time step q, $E_{inj,pv,j,q}$ values are the recorded 15-min energy generation values of customer j during time step q and $E_{tot,j}$ is the total yearly PV energy generation of customer j.

Once this is done, the 15-min CDFs of every user are aggregated, as graphically outlined in **Figure 5**, in order to create one reference CDF that can represent all PV owners in the specific feeder. For creating the CDF of each particular future PV owner, this reference CDF should be normalised in function of his annual PV generation. For existing PV owners, such information is usually available to the DSO even if the customer is not monitored by an SM device. In case of future PV nodes, such information is apparently not available since no PV unit is connected. Consequently, the reference CDF is normalised with the annual PV generation of an existing PV unit (in the feeder or in proximity) multiplied by a reference factor f, as explained in the following section.

5. PV hosting capacity computation

Practically, the algorithm starts with the probabilistic analysis of the current situation (existing PV units), by simulating a large number M of possible system states. It is important to note that although system states are based on 15-min resolution data, each one of them is considered as a possible instantaneous state of the system. Thus, the accuracy and reliability of the computation increase with the number of treated system states.

The probabilities $P_{\text{overvoltage}}$, $P_{\text{undervoltage}}$ and $P_{\text{unbalance}}$ are computed at every node, based on the analysis results. Compliance with the conditions set by (1) is verified for the whole feeder. Moreover, compliance with the maximum current capacity is verified in the entire feeder for all the studied system states. In case both conditions are respected, the algorithm increases the installed PV power at the future (specified by the user) PV nodes by the defined increase step. Therefore, let us consider an LV feeder that is simulated with a total number N of PV nodes. Some of the simulated N nodes may be currently existing PV nodes while the rest of them are the considered future PV nodes. If the total number of future PV nodes is equal to K ($K \leq N$), the new installed power at each future PV node i is computed as follows:

$$P_{rated,l,i} = P_{rated,l-1,i} + P_{step,i}\ ,\ i = 1 : K \text{ nodes} \tag{7}$$

where $P_{rated,l,i}$ is the new installed PV power at node i in the current configuration l that will be analysed by the algorithm (in step 5, Figure2), $P_{rated,l-1,i}$ is the installed PV power at node i that was analysed (and accepted in terms of impact on the technical constraints) in configuration l-1 and $P_{step,i}$ is the increase step (defined by the user for the respective node). A small P_{step}value (≈0.5-1kVA for residential or small commercial customers) is recommended so as to make a more precise computation. Note that in several countries, concerning residential and small-business customers, the maximum admissible installed power per distributed PV unit in the LV network is equal to 10kVA. In such cases, the condition $P_{rated,l,i} \leq 10$kVAshould be integrated in step 5 of the algorithm.

Once relation (6) is applied, the new installed PV power $P_{rated,l,i}$is defined at every new PV node before the algorithm performs the next "hosting capacity review" iteration (step 5, **Figure 2**). However, the reference CDF that represents the time-variability of generation at the new PV nodes needs to be scaled in function of $P_{rated,l}$ at each node. To do so, the reference CDF could be normalised in function of the annual PV generation of the PV unit. For existing PV owners, this information is usually available to the DSO even if the customer is not monitored by an

http://dx.doi.org/10.5772/62418

SM device. In this case, such information is not available since there are currently no PV units at the specified nodes. Consequently, the reference CDF is normalised with the annual PV generation of an existing PV unit (in the feeder or in proximity). Then, a reference factor f is introduced for scaling the normalised CDF in function of $P_{rated,l}$ The factor f_i is computed as follows:

$$f_i = P_{rated,l,i} / P_{rated,ref} \quad ,i = 1:K \tag{8}$$

where $P_{rated,ref}$ is the installed PV power of the existing PV unit that has been used to normalize the reference CDF.

Once the generation profiles have been set up for the future PV nodes, the algorithm repeats steps 2 and 3 for analysing the current configuration l. At this point, it is important to clarify that each "hosting capacity review" iteration l practically performs the power flow analysis of configuration l by applying a full MC simulation, similar to the one of step 2. This means that each "hosting capacity review" iteration l runs the same large number of MC iterations M that was analysed in step 2. Thus, in every iteration l, a very large number of system states is analysed (=M·96) so that the values of $P_{overvoltage}$, $P_{undervoltage}$ and $P_{unbalance}$ converge. Thanks to this procedure, the verification of compliance with equations (1) for each configuration l is assumed to be reliable. If the analysis of M system states, in configuration l, demonstrates that the operational constraints are not violated, the installed PV power is again increased at each future node. Then, the algorithm passes again to steps 4 and 5.

The described iterations stop as soon as the operational constraints are for the first time exceeded at least at one of the nodes. The PV size of some units could probably increase even more, given that the operational constraints at their PCC are not violated. However, this study treats the LV feeder as a whole since the violation of limits at one node is always affected by the energy flow at all nodes. The $P_{rated,l,i}$ that is applied in the last iteration l, which led to a violation of acceptable limits, is the one considered as the maximum admissible hosting capacity per node.

The aggregated PV hosting capacity of the feeder is computed by adding $P_{rated,l,i}$ (existing and new) along the feeder:

$$P_{rated,tot} = \sum_{i=1}^{N} P_{rated,l,i} \tag{10}$$

where N is the total number of PV nodes in the feeder. In order to make a more detailed computation, different increase steps could be applied per node in function of its position in the feeder. The voltage limits are usually more easily violated at the end of the line. Consequently, the PV power steps could be bigger for the nodes at the head of the line. However, this strategy could eventually result to an earlier (in terms of PV size) violation of the limits at the last nodes, which does not tally with a common welfare among end-users.

6. CASE STUDY: An LV feeder in Belgium

6.1. Description of the simulation

This section describes the application of the previously described analysis tool for computing the PV hosting capacity of an LV feeder in Flobecq. Flobecq is a municipal area in Belgium with high penetration of distributed PV generation (≈25% of Flobecq LV network customers) and large deployment of SM devices. Thanks to an official research fellowship between the local DSO and the authors' affiliation, the technical parameters of the feeder and SM datasets of the respective customers have been communicated strictly for research purposes. The datasets that were used in this case study cover a total period of one year (2013).

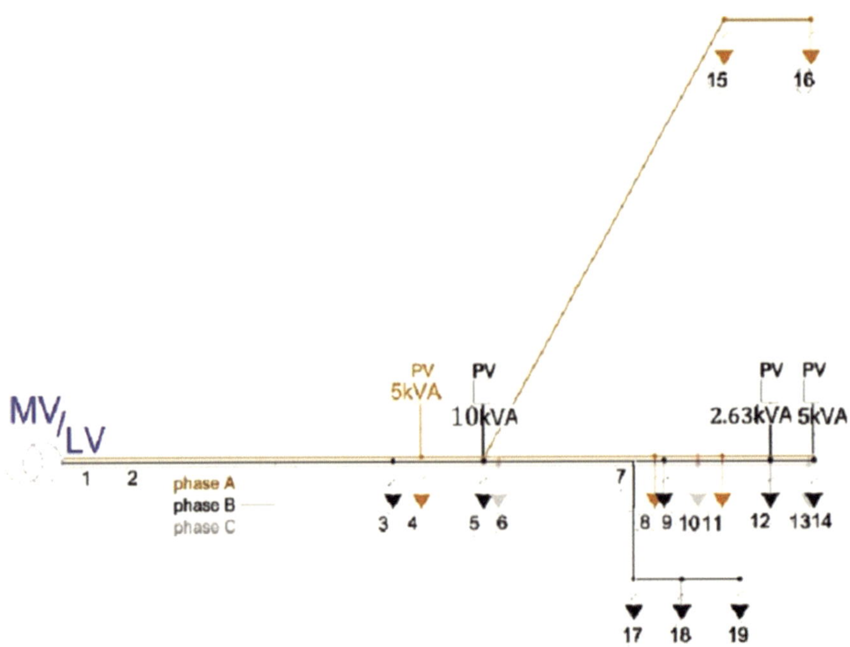

Figure 6. The simulated LV feeder of the power distribution network of Flobecq (conductors colour code as in IEC 60446 standard).

The topology of the simulated three-phase feeder is presented in **Figure 6**. Currently, four PV units are installed in the feeder which supplies a total of 16 residential customers. These PV units are located at nodes 4,5,12 and 14, by means of single-phase inverters, and their installed PV power is respectively 5kVA, 10kVA, 2.63kVA and 5kVA. A spatial correlation study had already been performed for the specific feeder and the generation profiles of the customers were proved to be entirely correlated [16]. This consideration is taken into account in this analysis, including the future PV nodes. Practically, this means that for every simulated system state, the randomly sampled probability value for defining PV generation is common for all PV units.

Concerning operational constraints, the ones of EN 50160 standard have been considered in the simulation. Therefore, compliance with the group of equations (1) has been verified for each system state, as far as voltage magnitude and unbalance are concerned. The maximum current capacity of the lines has been determined based on table [15]. The PV size increase step is defined equal to 1kVA and the power factor of all PV inverters is considered equal to 1, unless reactive power control is considered in the simulation.

A set of different scenarios have been simulated regarding the position and phase connection of future PV units as well as the action of voltage control schemes. The analysed scenarios are listed in **Table 1**.

Concerning scenarios A-D, only the on-off control scheme is considered, which is currently implemented by most DSOs in Europe. This control scheme enables a total cut-off of the PV unit (in most cases during 3 minutes) as soon as the voltage limit has been locally exceeded for a period longer than 10 minutes. This analysis considers each simulated state as instantaneous. Therefore, each violation of the 95-percentile limit of EN 50160 standard is counted in the probabilities even though in reality it might had lasted less than 10 minutes. This means that the computed maximum PV hosting capacity is possibly slightly lower than the one that the feeder can really support.

No	Description
A	12 new PV units at nodes 2, 3, 6, 7, 8, 10, 11, 13, 15, 17, 18, and 19. The PV units at nodes 8, 11, 17, 18, and 19 are connected to phase A, the PV unit at node 3 is connected to phase B and the PV units at nodes 2, 6, 7, 10, 13, and 15 are connected to phase C.
B	Similarly to scenario A but all new PV units connected to phase B, except from PV unit at node 15 that is connected to phase A.
C	3 new PV units connected to nodes 13, 18 and 19 (end of the line). All three new units connected to phase B.
D	1 new PV unit connected to node 13 (phase B).
E	Similarly to scenario A but considering 100-percentile and not 95-percentile operational limits. Practically the PV hosting capacity is not increased as soon as voltage and VUF limits are exceeded at least once in the feeder.
F	Similarly to scenario A but considering the action of three-phase damping control integrated in the new PV inverters. In this case, the new PV units need to be connected by means of three-phase PV inverters.
G	Similarly to scenario A but considering the action of reactive power control of (CEI 2012)

Table 1. The simulated PV hosting capacity scenarios.

The control scheme applied in scenario F is the three-phase damping control scheme which behaves resistively towards the negative- and zero-sequence voltage component, without modifying the injected power, so as to eliminate phase voltage unbalance [18]. This control scheme requires a three-phase PV inverter and it is very promising in terms of voltage magnitude and unbalance mitigation. It is actually implemented in a EU pilot program (FP7 INCREASE Project). The third control scheme is reactive power control in the way it is implemented in the Italian distribution system [20] concerning new PV units in the LV network.

6.2. Comparing with a deterministic approach

One of the main purposes of this study is to investigate, up to which extent, a probabilistic method based on customer-specific SM readings leads to a less restrictive computation of PV hosting capacity, compared to a deterministic approach. For this purpose, a deterministic approach has been implemented simulating worst case energy flow profiles. The load profiles of all customers and the PV generation profiles of existing PV units have been also based on SM recorded data. The deterministic steady state analysis has been conducted for scenarios A-D, F, G. Scenario E is not mentioned because, although SM readings are used, only 100-percentile limits are considered which means that probabilities are not accounted for in the computation of hosting capacity. Thus, this scenario is practically a deterministic scenario.

The following load/ PV generation profiles have been considered in the deterministic approach:

1. Maximum PV power per node (installed PV power) – Minimum recorded load per node; absolute values, irrespective of time coincidence among customers
2. Maximum PV power recorded in the feeder – Coincident PV generation/load values for the other nodes.
3. Minimum recorded load in the feeder during PV injection hours – Coincident PV generation/load values for the other nodes.

6.3. Results and discussion

The probabilistic hosting capacity review results are illustrated in **Figure 7** and analytically listed in **Table 2**. The aggregated maximum admissible PV hosting capacity in the feeder is presented for each individual scenario, considering separately the EN 50160 standard's voltage limits and the maximum current capacity of the lines. This separate presentation has been chosen because voltage limits are treated with a probabilistic approach in the EN 50160 standard while congestion risk is treated by each DSO with a different approach. Most of them apply a deterministic approach that considers an upper (100-percentile) current limit. The violation which forbade further increase of the PV hosting capacity is also presented and quantified for each scenario. The aggregated PV hosting capacity obtained with deterministic analysis is presented in **Figure 7** and **Table 3** for all treated scenarios (§5.2).

http://dx.doi.org/10.5772/62418

No	EN 50160 standard voltage limits' consideration		Maximum current capacity and voltage limits' consideration	
	Aggregated PV hosting capacity (kVA)	Violation	Aggregated PV hosting capacity (kVA)	Violation
A	154.63kVA (11kVA/ per new PV node + existing PV units)	-- $P_{overvoltage}$ at nodes 18 and 19 (phase (B)) resulted 5.7% and 6.4% respectively (>5%, which is the value accepted by the EN 50160 standard)	70.63kVA (4kVA/ per new PV node + existing PV units)	I_{max} of line 6–7: 13% deviation (13% higher than the maximum current capacity of the lines)
B	144.63kVA (10kVA/ per new PV node + existing PV units)	-- $P_{overvoltage}$ at nodes 13,14,15 (phase (C), resulted 5.4%, 6.16% and 6.18% respectively (>5%, which is the value accepted by the EN 50160 standard)	58.63kVA (3kVA/ per new PV node + existing PV units)	I_{max} of line 6-7: 50% deviation (50% higher than the maximum current capacity of the lines)
C	82.63kVA (20kVA/ per new PV node + existing PV units)	-- $P_{overvoltage}$ at node 19 (phase (B)) resulted 6.3% (>5%, which is the value accepted by the EN 50160 standard)	43.63kVA (7kVA/ per new PV node + existing PV units)	I_{max} of line 6–7: 10.5% deviation (10.5% higher than the maximum current capacity of the lines)
D	65.63kVA (43kVA/ per new PV node + existing PV units)	-- $P_{overvoltage}$ at node 19 (phase (B)) resulted 5.15% (>5%, which is the value accepted by the EN 50160 standard)	37.63kVA (15kVA/ per new PV node + existing PV units)	I_{max} of line 6–7: 6.2% deviation (6.2% higher than the maximum current capacity of the lines)
E	94.63kVA (6kVA/ per new PV node + existing PV units)	-- $P_{overvoltage}$ at nodes 13,14, (phase (C)), resulted 0.0001% in both cases (>0%, which is the condition in scenario E)	70.63kVA (4kVA/ per new PV node + existing PV units)	I_{max} of line 6–7: 13% deviation (13% higher than the maximum current capacity of the lines)
F	202.63kVA (15kVA/ per new PV node + existing PV units)	-- $P_{overvoltage}$ at nodes 2-19 (at all three phases) resulted from 5.5% to 28% (>5%, which is the value accepted by the EN 50160 standard) -- $P_{unbalance}$ at nodes 2-19 resulted from 10% to 32% (>5%, which is the value accepted by the EN 50160 standard)	70.63kVA (4kVA/ per new PV node + existing PV units)	I_{max} of line 6–7: 11% deviation (11% higher than the maximum current capacity of the lines)
G	154.63kVA	-- $P_{overvoltage}$ at node 19 (phase (B)) resulted 5.17% (>5%, which is the value accepted	70.63kVA	I_{max} of line 6–7: 11% deviation (11% higher than

No	EN 50160 standard voltage limits' consideration		Maximum current capacity and voltage limits' consideration	
	(11kVA/ per new PV node + existing PV units)	by the EN 50160 standard)	(4kVA/ per new PV node + existing PV units)	the maximum current capacity of the lines)

Table 2. Aggregated maximum PV hosting capacity and violated parameter for each simulated scenario (Probabilistic Simulation Tool).

No	Aggregated PV Hosting Capacity (kVA) (considering EN 50160 standard voltage limits)			
	Profile (I)	**Profile (II)**	**Profile (III)**	**Violation**
A	70.63kVA (4kVA/ per new PV node + existing PV units)	82.63kVA (5kVA/ per new PV node + existing PV units)	82.63kVA (5kVA/ per new PV node + existing PV units)	Overvoltage at all new PV nodes
B	58.63kVA (3kVA/ new PV node + existing PV units)	58.63kVA (3kVA/ new PV node + existing PV units)	58.63kVA (3kVA/ new PV node + existing PV units)	Overvoltage at all new PV nodes
C	43.63kVA (7kVA/ new PV node + existing PV units)	43.63kVA (7kVA/ new PV node + existing PV units)	43.63kVA (7kVA/ new PV node + existing PV units)	Overvoltage at all new PV nodes
D	37.63kVA (15kVA/ new PV node + existing PV units)	37.63kVA (15kVA/ new PV node + existing PV units)	37.63kVA (15kVA/ new PV node + existing PV units)	Overvoltage at all new PV nodes

Table 3. Aggregated maximum PV hosting capacity for each simulated scenario (Deterministic Approach).

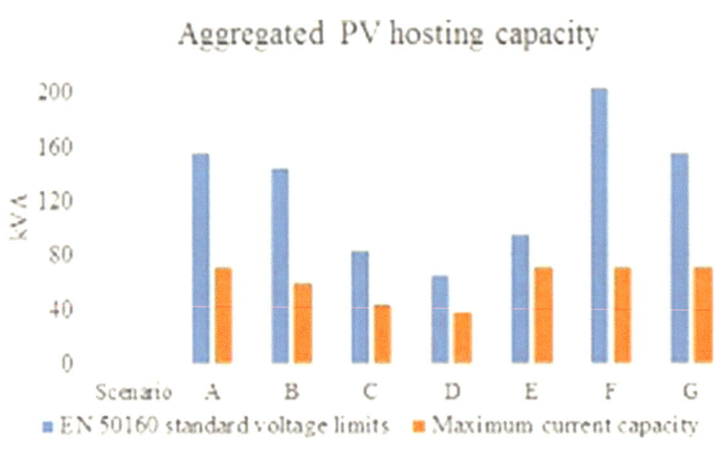

Figure 7. The computed aggregated PV hosting capacity of the feeder for scenarios A-G. The number of new PV units is also indicated.

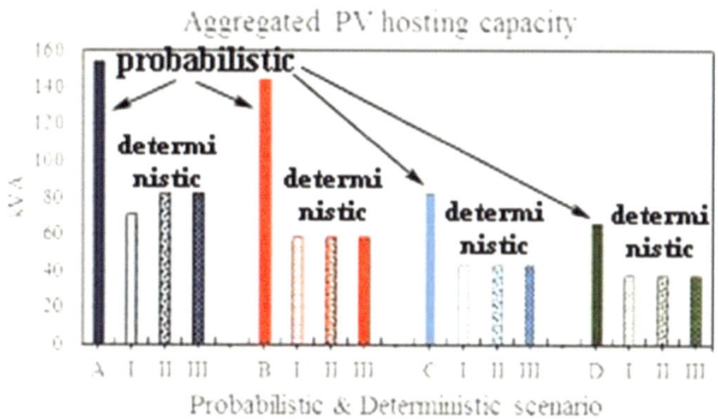

Figure 8. The computed aggregated PV hosting capacity of the feeder for scenarios A-D with the probabilistic and the deterministic approaches.

Firstly, one should note that the results considering the maximum current capacity coincide for the probabilistic and the deterministic computations since this metric is not addressed with probabilistic terms. In case the maximum value is exceeded at one point of the feeder during just one of the simulated states, the PV hosting capacity is not further increased in the respective simulation. Also, the result of scenario E (applying 100-percentile limits) is close to the ones of the deterministic scenarios A.I, A.II and A.III which analyse the same topology as scenario E but with a deterministic approach. Based on these remarks, one can reasonably assume that the probabilistic computation covers (samples and analyses) almost the whole range of possible system states, including the ones recorded in reality (the combination of coincidently recorded values) which are treated in the deterministic scenarios A.II and A.III.

However, accounting only for voltage violation, the restrictive condition of scenario E based on which voltage limits must never be exceeded (in none of the simulated states), results in a quite lower admissible PV hosting capacity compared to scenario A (same topology as scenario E). Basically, in scenario E, PV hosting capacity could not further increase because the computed $P_{\text{overvoltage}}$ resulted equal to 99.99% (>95% is the condition in EN 50160). Therefore, if the admissible PV hosting capacity does not exceed 94.63kVA, the operational limits will most probably never be violated in the feeder, based on the elaboration of the available historic data. Otherwise, if the admissible PV hosting capacity increases up to 154.63kVA, as in scenario A, voltage limits' violation will only take place in less than 5% of total system states. Therefore, even with such an increase of the aggregated PV hosting capacity, the temporary cut-offs of the PV units due to overvoltage will be very rare. Thus, scenario A takes advantage of the probabilistic character of EN 50160 standard (limits violation allowed during 5% of week time), which is not the case in scenario E or in the deterministic approach.

Figure 9 clearly demonstrates how the probabilistic consideration of overvoltage risk affects the computation. This figure shows the evolution of the CDF of phase voltage (B) at node 19 while the total installed PV power increases in scenario A. Based on the probabilistic analysis of the feeder, when total installed PV power increases by 144kVA (12kVA per new PV unit), phase voltage (B) at node 19 respects the defined limits in 94.6% of the simulated states.

However, EN 50160 defines that the limits should be respected in at least 95% of cases. Thus, the maximum PV power that can be added to the feeder, considering this configuration, is 132kVA (11kVA per new PV unit).

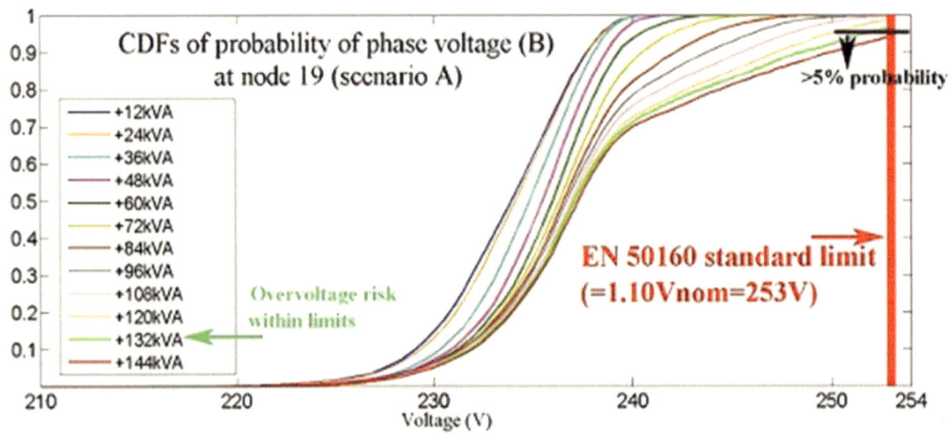

Figure 9. CDFs of probability for phase voltage (B) at node 19, for each increase step of the installed PV power in the feeder (scenario A).

The above arguments should be considered in a cost-benefit analysis (CBA) that compares costs for exceeding operational limits to losses due to an eventual penalty for low DER integration or loss of potential revenue for customers and energy utilities. Such considerations may allow a much more cost-effective PV integration strategy which also respects the applied standard's criteria. At the same time, the identification of critical points regarding congestion risk should also be considered. Both arguments highlight the usefulness of considering SM historic datasets in similar studies, so that critical points and probabilities are carefully mapped and quantified.

For highlighting the cost-effectiveness of deploying long-term measurements in the LV network and analysing it with a probabilistic approach, a more detailed computation of line losses in the feeder was performed for scenario A. Assuming that the computed maximum admissible PV power is installed (=154,63kVA if one considers only the voltage limits), the study focuses on the total energy losses along the lines of the feeder during hours of high PV injection in a typical day. The worst case approach considers only one system state which will more likely take place during hours with the highest PV injection. Based on the available historic data for the feeder, this period is between 12:00AM and 18:30PM on a typical July day. The sum of energy losses has been computed along the feeder for the considered period, for each simulated day. **Figure 10** illustrates the statistical distribution (CDF) of the computed line losses, obtained with the probabilistic approach.

http://dx.doi.org/10.5772/62418

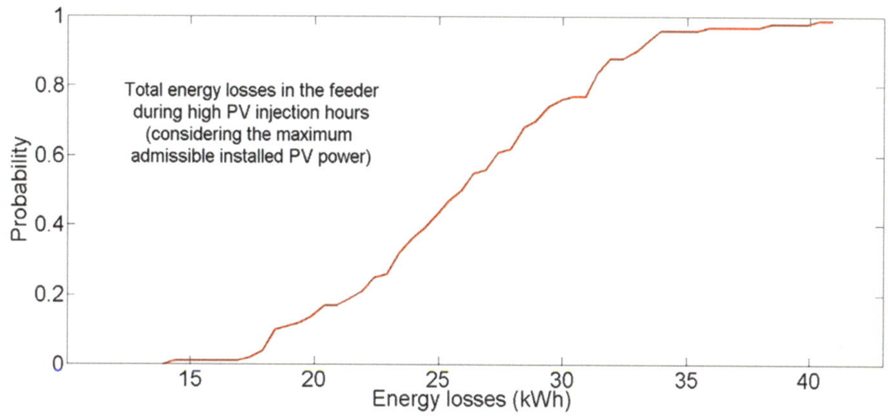

Figure 10. CDF of total energy losses in the feeder during high PV injection hours in a typical July day, considering the maximum admissible installed PV power (scenario A).

The probabilistic approach and the consideration of the SM measurements demonstrated that total energy losses in the feeder vary significantly, depending on the system state. Consequently, in 95% of the simulated days, total energy losses during high PV injection hours (12:00AM to 18:30PM) do not exceed 35kWh in a day. In the deterministic approach which assumes the worst case scenario taking place all along the high PV injection period, the respective energy losses result equal to 148kWh. This important difference is due to the fact that the probabilistic approach considers the extremely low frequency of worst case scenarios to take place simultaneously for all feeder users. Considering such probabilities, the DSO could manage a less conservative and more cost-effective long-term strategy.

Undoubtedly, the computed PV hosting capacity values depend on the load profiles of the customers that are located in the feeder. However, the results clearly indicate in relative terms, that smaller distributed PV units have a much smoother impact than the bigger ones concentrated in one small area of the feeder. This fact is demonstrated by the comparison of scenario A to scenarios C and D. Moreover, as previously mentioned, in several cases the maximum admissible installed power per PV unit connected to the LV network is equal to 10kVA. In such cases, scenarios C and D might not be appropriate based on the probabilistic simulation results. As a matter of fact, the admissible total installed power would have to limit to 32.63kVA (for scenario D) although the network would be able to support 37.63kVA. The difference between the PV hosting capacity computed with the probabilistic and the deterministic approach for these cases (considering only voltage limits) is not as big as for scenarios A and B. Indeed, in scenarios A and B, the volatile character and the extremely rare coincidence of worst case values for 12 units cannot be reliably represented by a deterministic model.

Regarding the distribution of units among phases, the comparison of scenarios A and B shows that the existing phase unbalance affected the computation. Indeed, the violated parameter in this case is voltage magnitude of phase (C) although all new PV units are connected to phase (B). Therefore, the unfair distribution of new PV units among phases did not directly affect $P_{unbalance}$ but it had an impact on the voltage magnitude of phase (C). Considering voltage limits, the aggregated PV hosting capacity for scenario B resulted equal to 144.63kVA if one considers the probabilistic character of EN 50160 standard's voltage limits. However, the connection of most new PV units at phase (B) resulted in very high current values so that the maximum current capacity was exceeded by 50%.

Scenario F demonstrated that the connection of new PV units by means of three-phase inverters integrating three-phase damping control can increase the aggregated hosting capacity by 36%, if the probabilistic character of EN 50160 standard's voltage limits is considered. Thanks to the resistive behaviour of this control scheme towards the zero- and negative-sequence voltage component, the deviation of voltage magnitude and unbalance becomes much smoother compared to the currently applied on-off control. Thus, the risk of exceeding the defined limits is reduced and a bigger share of PV generation can be integrated.

Based on the results of scenario G, reactive power control does not result in higher PV hosting capacity compared to scenario A (on-off control). Voltage profile in the feeder is however improved compared to scenario A. As a matter of fact, voltage limits are not violated in scenario G whereas the maximum current capacity limit is exceeded for the same amount of PV integration compared to scenario A.

In the first two cases (scenarios A and B), comparing the probabilistic simulation results to the respective ones of the deterministic approach, an important difference in the aggregated admissible hosting capacity is observed. At this point, it is important to mention that the violated parameter in the deterministic approaches is mainly the voltage magnitude and secondly the maximum current capacity of the lines. The deterministic approach led to 74-146% lower aggregated PV hosting capacity (compared to the one computed with the probabilistic approach) due to a violation that according to the probabilistic elaboration of the historic SM dataset took place for much less than 5% of the simulated system states. Indeed, based on **Figure 8**, the addition of 12 new PV units of 4kVA each (result of deterministic scenario A.I) generated an overvoltage risk that is lower than 1%.

In such cases, the probabilistic analysis demonstrates that the simulated worst case scenarios are extremely unlikely to happen in the studied feeder. However, such worst case approach is currently implemented by most DSOs when performing hosting capacity reviews. As a result, decisions for connecting new PV units in certain networks are very often extremely restrictive, with a big impact on the cost-efficiency of the network.

When it comes to scenarios C and D, one can note that if less but bigger size PV units are connected, the results of the probabilistic and the deterministic approach do not differ significantly. This result proves that in case of many distributed PV units, an approach that considers all worst case customers' profiles coinciding in time is mostly extreme. Deterministic approach cannot accurately simulate the volatile character of PV generation and the random loading parameters of residential and small commercial customers.

A general remark would concern the design strategy of distribution feeders like the studied one. The studied feeder currently hosts 22.63kVA of distributed PV generation and supplies 19 residential customers. The analysis of the current conditions (based on the historic SM datasets) demonstrated that both voltage violation risk and congestion risk are very low. Moreover, the above probabilistic load-flow analysis demonstrated that congestion and voltage problems will only appear if 48kVA and 132kVA respectively of distributed PV generation (scenario A) are further integrated. This remark highlights the cost-efficiency of designing distribution networks based on the most frequent system states or on well-studied future scenarios. This approach can lead to customised solutions and help to avoid over-dimensioning and costly initial investments for the DSO.

Based on the above analysis, certain renewable integration scenarios could increase to an important extent the self-sufficiency of feeders like the studied one. As a result, their dependency on big conventional power plants, connected at the transmission level, could be efficiently reduced. However, big conventional plants are important for maintaining grid stability. In a high DER integration scenario, without large and reactive storage facilities and/or flexibility services, the amount of RES should be carefully reviewed. To this end, costs induced by the use of grid services, including insurance against periods when it is not possible to consume own generated electricity, should be considered and reflected in the bill of generator owners [1]. Reliable feasibility studies and comprehensive CBAs are necessary for evaluating various strategies in the decision making process.

6.4. The role of customer-specific SM data in PV hosting capacity reviews in LV networks

The above analysis is based on the use of customer-specific SM energy flow readings. Various maximum PV hosting capacity scenarios have been analysed by applying a probabilistic steady state analysis of the feeder on a 15-min time scale, sampling from the available SM data. In this way, the real probability of worst case scenarios has been accounted for and as a result, a probabilistic view of several technical metrics has been enabled (voltage and current magnitudes, voltage phase unbalance, line losses). Even if DSO long-term planning is based on deterministic approach, the use of SM datasets can validate the considered worst case scenarios. Besides, the wide deployment of SM devices can offer other possibilities such as better coordination and control of technical parameters of the LV network as well as better visibility and adaptability to actual load and generation profiles of LV customers. The long-term planning of LV networks can become more customised to local conditions and therefore more cost-effective.

7. Conclusions

This chapter addresses the problem of determining the maximum PV hosting capacity that can be accommodated in a LV distribution feeder, while respecting local technical standards. To this purpose, a probabilistic simulation tool that uses as input customer-specific SM energy flow data and feeder-specific parameters is presented. A PV hosting capacity review for a municipal area in Belgium is used as a case study for evaluating the usefulness and reliability of the proposed tool. The study outcome demonstrates that it is to the interest of the DSO and of the grid users to deploy probabilistic analysis that considers the time-variability of load/PV generation, both in the time axis and between different customers' profiles. This variability of network state can be taken into account thanks to the deployment of long-term SM measurements in the studied network. Consequently, the further deployment of SM devices is strongly recommended for achieving a more cost-effective long-term planning and coordination of the LV network.

Acknowledgements

The authors of this chapter acknowledge the support of ORES, the DSO who manages the electricity and natural gas distribution grids in 193 communes in Wallonia (Belgium), in terms of funding and SM data supply, both indispensable elements for conducting this research work.

Nomenclature

LV	**Low Voltage**
PV	Photovoltaic
DER	Distributed Energy Resource
DSO	Distribution System Operator
SM	Smart Meter
MV	Medium Voltage
CDF	Cumulative Distribution Function
$P_{overvoltage}$	Probability of exceeding upper voltage limit
$P_{undervoltage}$	Probability of exceeding lower voltage limit
$P_{unbalance}$	Probability of exceeding voltage unbalance limit
$V_{i,j}$	Grid voltage at node i, phase j
V_{nom}	Nominal voltage in the feeder
$P_{rated,l,i}$	Installed PV power at node i, considered in iteration l in case it is a future PV node
$P_{rated,tot}$	Installed PV power in whole feeder
P_{step}	Increase step of the installed PV power at a node
f_i	Reference factor

Table 4. Nomenclature.

Appendix: The power flow algorithm

The sequence admittance matrix for the main line is constructed with (A.1), where 0, 1, 2 represent respectively the zero-, positive- and negative-sequence whereas the superscript Z stands for series admittance. The probabilistic algorithm samples the voltage value V_n at the slack node (secondary output of the MV/LV transformer), which remains fixed throughout the forward/backward sweep process as in (A.2) [21]–[22]. The bold type stands for vectors and the under bar stands for complex numbers. The length of vector $\underline{V}_{\text{initial,1 or 2 or 3}}$ is equal to the number N of the nodes in the feeder.

$$\left[Y_{012}^{Z}\right]=\begin{bmatrix} y_0^z & 0 & 0 \\ 0 & y_1^z & 0 \\ 0 & 0 & y_2^z \end{bmatrix} \tag{A.1}$$

$$\begin{aligned} \underline{V}_a^{\text{slack node}} &= V_n 0° \\ \underline{V}_b^{\text{slack node}} &= V_n 120° \\ \underline{V}_c^{\text{slack node}} &= V_n 240° \end{aligned} \tag{A.2}$$

$$\underline{\boldsymbol{V}}_{\text{initial},0} = 0\angle 0°, \underline{\boldsymbol{V}}_{\text{initial},1} = V_{\text{n}}\angle 0°, \underline{\boldsymbol{V}}_{\text{initial},2} = 0\angle 0° \tag{A.3}$$

in symmetrical components

The unbalanced laterals are solved with a forward-backward method in phase components, which gives the power injections $S_{\text{lat},x,i}$ per phase (x=a,b or c) at each lateral's root node i (the total transited power by the respective lateral). In the described case the unbalanced laterals are 1-phase lines; therefore, $S_{\text{lat},x,i}$ is computed per phase by a 1-phase algorithm. The computed $S_{\text{lat},x,i}$ replaces each unbalanced lateral at its root node in the main line. During the backward step, the phase currents due to the nodal loads are computed for the nodes of the main line with (A.4). $S_{\text{load},x,i}$ is computed for node i with (A.5) (if i is a simple node) and (A.6) (if i is the root node of lateral):

$$\underline{I}_{\text{load},x,i} = \left(\frac{\underline{S}_{\text{load},x,i}}{\underline{V}_{\text{initial},x,i}}\right)^* \left(x = a, b \text{ or } c \text{phase}\right) \tag{A.4}$$

$$\underline{S}_{\text{load},x,i} = \left(P_{x,i}\right) + j\left(Q_{x,i}\right) \tag{A.5}$$

$$\underline{S}_{\mathrm{load},x,i} = \left(P_{x,i}\right) + \mathrm{j}\left(Q_{x,i}\right) + \underline{S}_{\mathrm{lat},x,i} \tag{A.6}$$

Active power values $P_{x,i}$ are defined by the MC algorithm, whereas reactive values $Q_{x,i}$ are calculated with constant values $\cos\varphi_{\mathrm{load}}<1$ (in case the node consumes energy from the network) and $\cos\varphi_{\mathrm{inj}}=1$ (in case the node injects energy in the n). Once the $\underline{\boldsymbol{I}}_{\mathrm{load,abc}}$ (phase components) matrix is constructed, it is transformed by means of the Fortescue transformation into the respective $\underline{\boldsymbol{I}}_{\mathrm{load,012}}$ (sequence components) matrix. The specified nodal loads for the positive sequence $\underline{\boldsymbol{S}}_{\mathrm{load,1}}$ are calculated with (A.7). At this point, the positive sequence nodal voltages V_1 are computed by applying the 1-phase load flow algorithm of [23]. The negative and zero sequence voltages are computed by solving the linear systems (A.8):

$$\underline{\boldsymbol{S}}_{\mathrm{load},1} = \underline{\boldsymbol{V}}_{\mathrm{initial},1} \cdot \left(\underline{\boldsymbol{I}}_{\mathrm{load},1}\right) \tag{A.7}$$

$$\left[y_0^Z\right]\underline{\boldsymbol{V}}_0 = \underline{\boldsymbol{I}}_{\mathrm{load},0} \; \& \left[y_2^Z\right]\underline{\boldsymbol{V}}_2 = \underline{\boldsymbol{I}}_{\mathrm{load},2} \tag{A.8}$$

Thus, the $\boldsymbol{V}_{012}$ matrix is constructed and transformed into the respective $\boldsymbol{V}_{\mathrm{abc}}$ matrix for the main line. The values of V_x (x=a, b or c) at the root nodes of the laterals are updated and the 1-phase algorithm is once again applied to compute the new nodal voltages of the laterals' nodes. At this point, the first forward/backward step of this hybrid power flow method is completed and the nodal voltages of the whole network (main line and laterals) are updated with the new values (A.9). After each iteration (l) of the power flow algorithm, the convergence error E is calculated for phase x at node i with (A.10) where N is the number of nodes of the whole feeder. As soon as the error $E_{x,i}$ becomes smaller than a given tolerance, the algorithm stops after iteration l is completed and the last computed $|\boldsymbol{V}_{\mathrm{abc}}|$ is compared to the operational limits.

$$\underline{\boldsymbol{V}}_{\mathrm{initial},x} = \underline{\boldsymbol{V}}_x \quad \left(x = \mathrm{a, b\ or\ c phase}\right) \tag{A.9}$$

$$E_{x,i} = \left|V_{x,i}^{(l)} - V_{x,i}^{(l-1)}\right| \left(x = \mathrm{a,b,c\ and\ } i = 1:\mathrm{N}\right) \tag{A.10}$$

Author details

Vasiliki Klonari*, Jean-François Toubeau*, Jacques Lobry and François Vallée*

*Address all correspondence to: Vasiliki.KLONARI@umons.ac.be, jean-francois.toubeau@umons.ac.be and Francois.VALLEE@umons.ac.be

Electrical Engineering Department, University of Mons, Mons, Belgium

References

[1] EDSO, "European Distribution System Operators for Smart Grids Adapting distribution network tariffs to a decentralised energy future," 2015.

[2] EN50160, "Voltage characteristics of electricity supplied by public electricity networks," 2012.

[3] M. B.-P. Antoni Klajn, "Application Note Standard EN50160 Voltage Characteristics of Electricity Supplied by Public Electricity Networks," March, 2013.

[4] Solar City Grid Engineering, "Integrated Distribution Planning: A holistic approach to meeting grid needs and expanding customer choice by unlocking the benefits of distributed energy resources," 2015.

[5] J. C. Hernandez, F. J. Ruiz-Rodriguez, and F. Jurado, "Technical impact of photovoltaic-distributed generation on radial distribution systems: Stochastic simulations for a feeder in Spain," *Int. J. Electr. Power Energy Syst.*, vol. 50, no. 1, pp. 25–32, 2013.

[6] F. J. Ruiz-Rodriguez, J. C. Hernández, and F. Jurado, "Probabilistic load flow for photovoltaic distributed generation using the Cornish-Fisher expansion," *Electr. Power Syst. Res.*, vol. 89, pp. 129–138, 2012.

[7] P. Zhang, W. Li, S. Li, Y. Wang, and W. Xiao, "Reliability assessment of photovoltaic power systems: Review of current status and future perspectives," *Applied Energy*, vol. 104. pp. 822–833, 2013.

[8] R. Billinton and Bagen, "Generating capacity adequacy evaluation of small stand-alone power systems containing solar energy," *Reliab. Eng. Syst. Saf.*, vol. 91, no. 4, pp. 438–443, 2006.

[9] Bollen M.H.J. and F. Hassan, *Integration of Distributed Generation in the Power System*, IEEE Press. Wiley, 2011.

[10] R. Billinton and R. Karki, "Reliability/cost implications of utilizing photovoltaics in small isolated power systems," *Reliab. Eng. Syst. Saf.*, vol. 79, no. 1, pp. 11–16, 2003.

[11] European Photovoltaic Industry Association, "Connecting the Sun. Solar Photovoltaics on the Road to Large-Scale Grid Integration," 2012.

[12] European Photovoltaic Industry Association, "Global Market Outlook for Photovoltaics," 2014.

[13] F. Vallee, V. Klonari, T. Lisiecki, O. Durieux, F. Moiny, and J. Lobry, "Development of a probabilistic tool using Monte Carlo simulation and smart meters measurements for the long term analysis of low voltage distribution grids with photovoltaic generation," *Int. J. Electr. Power Energy Syst.*, vol. 53, pp. 468–477, 2013.

[14] V.Klonari, J-F.Toubeau, Z. De Greve, O.Durieux, J.Lobry, F.Vallee, "Probabilistic Simulation Framework of Balanced and Unbalanced Low Voltage Networks," *Int. J. Electr. Power Energy Syst.* (Article in Press), 2016.

[15] IEC, "IEC 60364-5-52 Table A5210." .

[16] F. Vallée, F. Moutier, V. Klonari, J. Toubeau, and J. Lobry, "Clustering of Photovoltaic Generation for the Consideration of Time Changing Geographical Correlation in Probabilistic Analysis of Low Voltage Distribution Systems," in *5th Solar Integration Workshop*, 2015.

[17] S. Lefebvre, "Contribution à l'utilisation d'un outil d'analyse technico-économique de réseaux de distribution Basse Tension en l'absence de compteurs électriques intelligents,", UMONS, 2015.

[18] B. Meersman, B. Renders, L. Degroote, T. Vandoorn, and L. Vandevelde, "Three-phase inverter-connected DG-units and voltage unbalance," *Electr. Power Syst. Res.*, vol. 81, no. 4, pp. 899–906, 2011.

[19] FP7 INCREASE Project, "http://www.project-increase.eu/." .

[20] C.E.I. 0–21, "Reference technical rules for the connection of active and passive users to the LV electrical utilities," 2012.

[21] M. Abdel-Akher, K. M. Nor, and A. H. Abdul-Rashid, "Development of unbalanced three-phase distribution power flow analysis using sequence and phase components," in *2008 12th International Middle East Power System Conference, MEPCON 2008*, 2008, pp. 406–411.

[22] M. Abdel Akher, K. M. Nor, and a. H. a Rashid, "Improved three-phase power-flow methods using sequence components," *IEEE Trans. Power Syst.*, vol. 20, no. 3, pp. 1389–1397, 2005.

[23] I. T. Papaioannou, A. Purvins, and E. Tzimas, "Demand shifting analysis at high penetration of distributed generation in low voltage grids," *Int. J. Electr. Power Energy Syst.*, vol. 44, no. 1, pp. 540–546, 2013.